INSECT
POPULATION ECOLOGY

INSECT POPULATION ECOLOGY

an analytical approach

G.C.VARLEY

Hope Department of Zoology
(Entomology) Oxford

G.R.GRADWELL

Department of Forestry, Oxford

M.P.HASSELL

Department of Zoology and
Applied Entomology
Imperial College of Science
and Technology, London

UNIVERSITY OF CALIFORNIA PRESS
BERKELEY AND LOS ANGELES 1974

UNIVERSITY OF CALIFORNIA PRESS
Berkeley and Los Angeles, California

ISBN: 0-520-02667-5
Library of Congress Catalog Card Number: 73-89367

Printed in Great Britain

CONTENTS

PREFACE

This book is written mainly for university students and research workers. The objective is to give the biologist a more critical approach to the interpretation of population figures than will be found in most current texts of general ecology.

It was in 1969 when we were all concerned with teaching a course on insect population dynamics to students at Oxford that we decided to collaborate and prepare this book. Now we teach in three departments in two universities. This separation has helped to delay the book, but has enabled us to include a lot of new and valuable facts and ideas, as will be seen from the many citations in the bibliography of papers published in the last four years.

We wish to thank the many students with whom we have discussed the problems in this book. They have forced us to clarify our ideas, reinterpret a number of the basic ecological papers, and to simplify our terminology.

ACKNOWLEDGEMENTS

The facts and figures on which this book is based come from many sources and we are grateful especially to the following for permission to copy or simplify their graphs or illustrations. Dr Henry Bess for Fig. 3.11, Dr C. P. Clausen and the US Department of Agriculture for Table 9.1A, Dr A. C. Crombie and the Royal Society of London for Figs 2.1, 3.3 and 3.5, Dr D. J. Lewis and the Royal Society for Fig. 5.4, Professor A. C. Hodson for Fig. 8.4 and part of Fig. 8.5, for the rest of which we are indebted to Dr W. Baltensweiler. Dr R. F. Morris kindly allowed us to combine two of his figures to make Fig. 8.7, Professor D. W. Wright and the Association of Applied Biologists allowed us to print Fig. 9.9, and Mrs Phyllis Nicholson kindly permitted us to use some graphs published by the late Dr A. J. Nicholson which appear in Figs. 2.13 and 2.15A.

Some of the illustrations have already appeared in our own publications and we wish to thank the Royal Society of Arts for permission to re-use the pictures in Fig. 9.2 and Figs. 9.3A and c, the Tall Timbers Research Station for the use of Figs 7.1, 7.2 and 7.5, and the Editor of *Nature* for Figs 4.9 and 4.10. The British Ecological Society has kindly allowed us to re-use the graphs which appear as Figs. 4.8 and 8.2 and to use the late Dr Davidson's graph in Fig. 5.2.

Most of the drawings and inset sketches have been specially prepared by the senior author or by Mrs Rosemary Wise, to whom we are specially grateful for Figs 8.1 and 9.4A and c. Derek Whiteley drew Fig. 9.1 which is reproduced by permission of *The Countryman*.

We wish to thank Dr D. J. Rogers who kindly read the text and made various helpful suggestions, and Professor M. J. Way for constructive comments on Chapter 9.

CHAPTER 1

EXPRESSING POPULATION CHANGES

1.1 Synopsis

Census figures for insect populations can be represented in a number of different ways; we distinguish total population curves, partial population curves, generation histograms and generation curves. These can be plotted either on logarithmic or arithmetical scales.

With full census information, population changes can be assigned either to reproduction or to mortality from some identifiable causes. Mortality can be expressed either numerically or as percentages of survival or mortality, but is most conveniently expressed as the k-value; this is a measure of the 'killing power' of a particular mortality factor on a logarithmic scale.

1.2 Introduction

Population studies are rapidly gaining in importance partly because of the human population explosion. The same basic principles probably govern the populations of man and of other animals, but insects provide exceptionally simple material for field studies of population problems. They are also convenient for experimental study in the laboratory, where they can be cultured cheaply for many generations in a short time. Some insects which are serious pests of agricultural or forest crops have been counted and studied in detail so that the figures for many consecutive generations can be compared and analysed. Other important insects like lice, mosquitoes, and tsetse flies which transmit human diseases are harder to count. Ways of reducing their numbers have been sought for most of the last century with varying success, but progress in the understanding of their status has been slow. We have found few examples from medical entomology to illustrate the principles of population dynamics. Too often only the easy things

1

have been measured, and the interpretation of the figures demands observations which have not yet been made.

Research workers who already have some familiarity with the literature will realize how much our understanding of any basic ecological problem depends on the choice of an easy animal to study and on the planning of the census work or experimental study. Ideally research workers should fully understand both the theoretical background and the methods they will use for the analysis of their figures at the time the work is being planned so that no vital measurement is omitted. Of course this ideal is hard to realize, especially when methods are changing; but without very careful planning the results of ecological work may be difficult or impossible to interpret.

In this short book we make no attempt to give comprehensive coverage of the literature. We have concentrated on the clearest and simplest examples either from laboratory studies or from field studies of insects which have a single generation in the year. These have been easier to study than animals like aphids, with many generations in the year, or animals like locusts which migrate. Census figures for such migrants defy detailed interpretation at the present time.

So this book is concerned with those census figures which we are beginning to understand. Scientific understanding implies that the principles are clear and that their application provides a satisfactory explanation of the observations. The basic theories about populations were published long before we had any good census figures, and are commonly expressed in mathematical terms. To biologists who think in visual terms mathematics is like a foreign language: but the assumptions mathematicians use are normally extremely simple and can be presented as linear graphs when the right co-ordinates are used. So we have expressed census figures in ways which enable us to test the theories. Some theories are confirmed, but others give such a poor match to observation that they must be rejected.

Statistical treatment has been kept to the bare minimum because most of the points we wished to make are reasonably clear without it. However, many numerical exercises are included which are so graded that biologists without detailed mathematical knowledge can teach themselves how to analyse census figures, stage by stage.

At the end of this chapter we show some of the ways in which population figures can be expressed. In Chapters 2 and 3 we discuss the effects of competition between individuals and between species for their needs such as food or space. Then we discuss in Chapter 4 how predators and parasites behave in their search for their specific food requirements, and in Chapter 5 the effects of weather and climate. In the last four chapters we consider ways of handling the complexities of real field problems. Chapter 6 is concerned with preparing life tables for the simple case of an insect with an annual life cycle. Even a short series of life tables leads to a deeper understanding of how the animal and its environment interact. Chapter 7 gives a simple account of our analysis of a longer series of census figures for the winter moth and shows how far we can model the population changes and 'explain' them. In contrast, Chapter 8 considers how far outbreaks of forest insect pests can be understood, even when we have incomplete life table data. Lastly we consider the practice of biological control of agricultural and forest pests.

We understand the populations of man and other animals less well than the populations described in this book because the necessary facts are not yet available for them. If it is important enough, and the difficulties are faced and overcome, then ways will be found to understand other kinds of populations and manage them if necessary. In this book we try to lay firm foundations.

1.3   Population curves

Before discussing the real differences between the ways in which populations change in the field or in the laboratory, we must first eliminate the apparent differences which arise from the diverse methods which are used to present census figures.

Suppose that we had complete information available about the changes in numbers of an insect with an annual life cycle, as illustrated for a hypothetical species in Fig. 1.1A. There is a short breeding season during which the females lay their eggs, after which the adults die. From the eggs hatch the larvae, which pass through three stages during which they feed and grow. The long pupal stage leads to the adult stage in the next season.

The *total population curve* represents the total number of individuals of all stages, or their population density, plotted against time. Within the life-span of a single insect we require many observations and the numbers of eggs, larvae, pupae and adults are summed for each count.

There is, in each generation, a large peak of numbers in the breeding season when the egg stage predominates numerically.

The *partial population curves* represent separately the numbers of eggs, the numbers of larvae in each different stage, the numbers of pupae and the numbers of adults present. You will notice in our example that there is a time when *all* individuals are pupae, so that at this season, the pupal population curve and the total population curve coincide. Notice that the peaks of successive stages tend to be smaller and smaller, because some insects die in each stage. When adults and eggs are present together and when there are larvae of different stages present at the same time, the peak number of any stage is considerably less than the total population. No single census count in Fig. 1.1A will provide directly a figure for the total number of insects entering a given stage of the life cycle. When we need these figures—as we do to prepare a life-table (Chapter 6)—they must be obtained indirectly from a series of census counts. For example, let us suppose that we wish to find the number of adults emerging in a generation from a partial population curve for adults. The simplest method is to calculate the area beneath the adult curve, which represents 'adult days', and divide this by the mean adult life span. This method is only appropriate if any mortality of adults occurs entirely at the end of the stage. If mortality is fairly constant throughout the stage, the method of Richards & Waloff (1954) is likely to be more accurate. Several of these techniques have been well reviewed by Southwood (1966).

1.4   Generation histograms and generation curves

After we have estimated population density or the number of individuals passing through some particular stage from census data for a number of different generations, we can plot these figures against the particular generation number. Using the population curves in Fig. 1.1A we have plotted for each year histograms for the first stage larva and the adult insects in Fig. 1.1B. The figure used for the adult generation histogram is the peak of the *cumulative adult population* curve of each generation. Data of this kind are logically presented as a histogram. In practice the information is often presented as a *generation curve*, in which lines join successive points. This is convenient but illogical, because the line does not correctly represent the population at times intermediate between the points as would be the case in a population curve

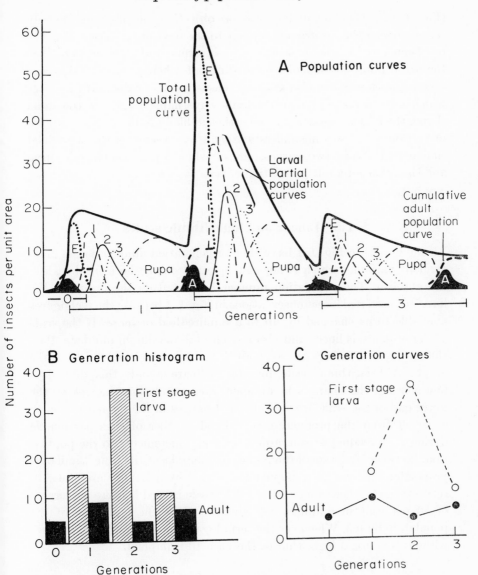

Fig. 1.1 Hypothetical population figures for three insect generations.
A The **partial population curves** for the egg E, the larval stages 1–3,
the pupa and the adult A are summed to give the **total population
curve**. The asymptote of the **cumulative adult population curve**
shows the total adult emergence for each generation.
B Some of the same figures plotted as a **generation histogram**.
C The same figures as in B shown as **generation curves**.

(Fig. 1.1A). The line in fact has no objective meaning, but merely serves to lead the eye from one point to the next in the series. Fig. 1.1 has been provided here to define these terms, and to emphasize that the same information looks quite different when plotted in different ways. The converse is also true—if census results or theoretical curves which have been plotted in different ways happen to have the same shape, their basic causation is nevertheless likely to differ. We shall see in Section 2.6 how a misunderstanding arose because of the superficial similarity between two curves, one of which was a generation curve and the other an adult population curve.

1.5   Linear and logarithmic scales

All the graphs in Fig. 1.1 have time or generation number as the independent variable, which is normally plotted on the horizontal scale— the abscissa. The population number is plotted on the ordinate. We have not used real insect observations in Fig. 1.1 because the real figures available to us changed by up to a hundredfold or more. If the scale of the ordinate is linear and also includes the maximum numbers, then the minimum numbers are so small that they cannot be read from the graph. A logarithmic scale for the ordinate avoids this difficulty; however big the changes in numbers are, population changes at the lower end of the scale are still visible. Changes in the population, such as result from the production of a fixed number of eggs per female during the breeding season, appear as equal increments to the population, however big or small the population may be at the time. Similarly, proportional decreases in population caused by mortality are easier to relate to each other when a logarithmic scale is used because a decrease of 50 per cent in numbers will always reduce the logarithm of the population by 0·3. Whenever the term logarithm or log is used, as here, without a suffix, a logarithm to the base 10 is implied.

1.6   Mortality and survival

We can illustrate simply the different ways in which mortality and survival can be expressed with the hypothetical figures in Table 1.1. Line 1 of the table shows the changes in numbers during a single generation. The thousand eggs hatch to produce only 100 small larvae,

50 survive to become large larvae, 20 reach the pupal stage, and 10 reach the adult stage.

Line 2 of the table shows the number which die in each interval between the counts and the sum (990) is the total number of individuals that die before the adult stage is reached. In line 3 are figures for the percentage mortality calculated in relation to the original egg number. When summed, these percentages equal 99 per cent—only 1 per cent of the eggs have produced adults. The figures showing the mortality in each stage as a percentage of the number of individuals alive at the beginning of the stage we term the *successive percentage mortality* (line 4). It is sometimes known as the 'apparent mortality' in contrast to the 'real mortality' in line 3. When mortality is calculated as successive percentages, the sum of the percentages has no meaning. Line 5 shows the percentage which survive at each stage and this is expressed as a decimal fraction in line 6. The fraction surviving is useful in a variety of calculations about populations; the product of the values of fractions surviving gives us the generation survival of $0 \cdot 01$. If we convert the population figures to logarithms as in line 7, then the effects of the mortality factors can be expressed logarithmically as their *killing power*, or k-value, which is the difference between the logarithms of the population before and after the mortality acts. So when the egg population ($\log N_E = 3$) changes to a population of small larvae ($\log N_{L1} = 2$) the k-value of the mortality factors concerned is $3 - 2 = 1$. For the pupal mortality, the k-value is $0 \cdot 3$. This method of expressing successively acting mortalities has the advantage that the k-values can be added up; and since they act in sequence their sum equals the generation mortality K ($K = 2 \cdot 0$ in Table 1.1). Inspection of the last two lines of Table 1.1 shows how convenient and easy it is to express populations by their logarithms and the effects of a mortality by the k-value, because the new population is obtained by subtraction of the k-value from the logarithm of the preceding population.

Reproductive rates are conveniently expressed logarithmically. If Table 1.1 represents a stable population, the adults must have laid 200 eggs per female (with the sexes in equal proportion a mean egg capacity of 100); the log fecundity then has the value of 2. On this system, when a population is in balance the logarithmic increase $\log F$ and the generation mortality K must be equal. If the mean egg production per adult varies from generation to generation, then this change can also be expressed as a k-value (k_0) as in Chapter 7.

Table 1.1 Ways of expressing population change and age-specific mortality when stages of the insect do not overlap.

Line		Eggs N_E	Small larvae N_{L1}	Large larvae N_{L2}	Pupae N_P	Adults N_A	
1	Population	1000	100	50	20	10	
2	Number dying in interval	900 +	50 +	30 +	10		Sum: 990 dead
3	% Mortality	90 +	5 +	3 +	1		Sum: 99% mortality
4	Successive % mortality	90	50	60	50		
5	Successive % survival	10	50	40	50		
6	Fraction surviving	0·1 ×	0·5 ×	0·4 ×	0·5 ×	1·0	Product = 0·01 survival
7	Log population	3·0	2·0	1·7	1·3	1·0	
8	k-value	1·0 +	0·3 +	0·4 +	0·3 +		Sum: $K = 2·0$

1.7 Overlapping generations and continuous breeding

The simple ways of expressing numbers and population changes which we have illustrated in this introduction are applicable to many insects and other animals but not to all. The most intractable populations are those with overlapping generations, where a lot of stages live together. For such insects as aphids, detailed census figures are very laborious to obtain and very hard to interpret. In this elementary book we shall confine ourselves largely to cases which are relatively simple to interpret because the amount of overlap between stages is small.

CHAPTER 2

DENSITY DEPENDENT PROCESSES AFFECTING CULTURES OF SINGLE SPECIES

2.1 Synopsis

When resources are in short supply, competition between individuals reduces their reproductive rate or their survival. The effects of competition become greater with crowding and are therefore density dependent. The effects can conveniently be isolated for experimental study in insect cultures under constant conditions where food supplies are renewed. For insects with overlapping generations the observations have been described in terms of the Verhulst–Pearl equation, which gives a sigmoid or logistic curve. When generations do not overlap, or when old and young individuals are not equal, changes in the structure of the population become important, and the logistic curve describes the observations inadequately.

Realistic population models must include age-specific mortality. If an insect population with synchronized generations suffers only a density dependent mortality, the subsequent population changes depend on the effective rate of increase and on whether competition is weak, moderate or severe. Moderate competition which undercompensates or only slightly overcompensates for population change may regulate the population at a stable level. In 'Scramble' competition some of the scarce resource is wasted and the resulting overcompensation can lead to a separation of generations and violent population change from one generation to the next. Apparently simple experiments on competition create situations where there are complex changes in both insect behaviour and insect numbers.

10

2.2 Intraspecific competition

Many species of insect can be reared in small containers. By providing a regular food supply, or by renewing the food at regular intervals, the cultures can be maintained almost indefinitely. If ways can be found to count the numbers present, we can observe how the population changes in the course of time under uniform conditions. These experiments are usually performed in an incubator which maintains a constant humidity

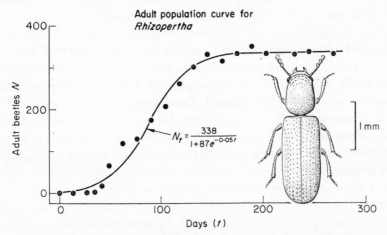

Fig. 2.1 Number of adult beetles in a culture started with a single pair of *Rhizopertha dominica* in 10 g of wheat grains, sifted and made up to 10 g every week. Inset: adult beetle. Data from Crombie (1945).

and temperature so the conditions are extremely artificial; but the experiments are in the tradition of scientific investigation because they permit the control of all variables except the population number. Intraspecific competition for food and for space become more important under these restricted conditions and their effects can be studied in comparative isolation.

Here we use the term **competition** in a precise way. Bakker (1961, 1969) defined competition as a 'manifestation of the struggle for existence in which two or more organisms of the same or of different species exert a disadvantageous influence upon each other because their more or less active demands exceed the immediate supply of their common resources'. Similar definitions have been given by Solomon (1949), Andrewartha & Birch (1954) and Milne (1957). All include the idea of *harm* being done to a number of animals striving for a *resource* which is in *limited supply*. We particularly like Bakker's definition because, by including the word 'manifestation' he draws attention to the fact that it is just as important to be able to *measure* the effect of competition as it is to know its *cause*. To assess the effect of competition we must (1) measure the changes in the supply of the resources, (2) measure the number of individuals competing in the population, and (3) assess the disadvantageous influences which may show either as a reduction in the number or proportion which survive, or as a reduction in growth rate, adult weight or reproductive capacity.

In this chapter we are concerned only with competition between individuals of the same species. To take a simple example first, Crombie (1945) began cultures of the lesser grain borer, a small Bostrychid beetle *Rhizopertha dominica*, with single pairs of beetles in 10 g of wheat grains (approximately 200 grains). These grains had been 'cracked' by slight compression. The female lays her eggs only in such cracks. Every week the grains were sieved off, their weight made up to 10 g with freshly cracked grains and the powdery faecal matter was rejected. This procedure kept the food resource approximately constant. Beetle eggs, larvae and pupae remained hidden in the grains and were not counted, but live and dead adult beetles were counted every two weeks.

The result of the experiment (Fig. 2.1) seem simple—the population built up steadily until it reached a stable level with a mean value of 338 beetles.

2.3   Theoretical background

If population growth is regarded as a continuous process, a mathematical model of population change can be derived from simple differential equations. If the population number at time t is N_t, then the population's instantaneous growth rate $(\mathrm{d}N/\mathrm{d}t)$ is given by:

$$\frac{\mathrm{d}N}{\mathrm{d}t} = r_m N \tag{2.1}$$

where r_m is the *intrinsic rate of natural increase* of the population (the maximum rate of increase = birth rate − death rate under fixed conditions of temperature and humidity). If the population number at time t_0 is N_0, then by integration the population at time t is given by

$$N_t = N_0 \exp(r_m t) \tag{2.2}$$

In natural (Napierian) logarithms this becomes

$$\log_e N_t = \log_e N_0 + r_m t \qquad (2.3)$$

or in logarithms to the base 10.

$$\log N_t = \log N_0 + r_m t (\log e) \qquad (2.4)$$

Curves based on these equations are shown as curve (a) in Figs. 2.2A and 2.2B on arithmetical and logarithmic scales respectively.

Malthus in 1797 followed this line of thought in relation to human population. He explained that the increase could not continue indefinitely and that 'misery and vice' must eventually limit the rise. Figure 2.1 showed that *Rhizopertha* population ceased to increase once it had reached a level a little above 300. Verhulst (1838) and Pearl & Reed (1920) independently expressed this idea mathematically in what is now known as the logistic or Verhulst–Pearl equation. They assumed that the *actual* rate of increase per individual as opposed to the intrinsic rate (r_m) which is a constant) is reduced as the population (N) rises to a stable upper limit (κ) which is the carrying capacity.

$$\frac{dN}{dt} = r_m N \left(\frac{\kappa - N}{\kappa} \right) \qquad (2.5)$$

It is the addition of the term $(\kappa - N)/\kappa$ in this equation that makes the logistic curve depart from the exponential growth curve as shown in Fig. 2.2. As the population increases, the rate of increase (r), where $r = r_m(\kappa - N)/\kappa$, is progressively reduced and it approaches zero (birth rate = death rate) when the population has reached the value of κ.

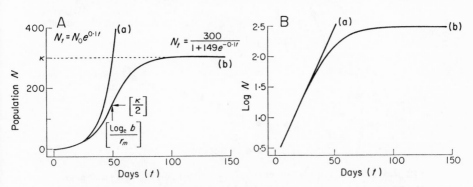

Fig. 2.2 Calculated exponential and logistic population growth curves.
A Population N plotted against time.
B Population plotted as log N against time.

This is clearly shown in Fig. 2.2B. The important biological implication of this population model is that there is *feedback* between the size of the population and the rate at which the population increases.

The logistic curves in Figs. 2.1 and 2.2 have been fitted using the integral form of equation (2.5)

$$N_t = \frac{\kappa}{1 + b \exp(-r_m t)} \qquad (2.6)$$

The constant b is related to the point of inflection of the curve (t') on the time axis

$$t' = \frac{\log_e b}{r_m} \qquad (2.7)$$

The predicted population size at this time (N') is always half of the value of κ, ($N' = \kappa/2$). These co-ordinates are shown in Fig. 2.2A.

Crombie (1945) obtained a very satisfactory agreement between a logistic curve and his numbers for *Rhizopertha* (Fig. 2.1), but he was wrong to claim that the biological assumptions on which the logistic

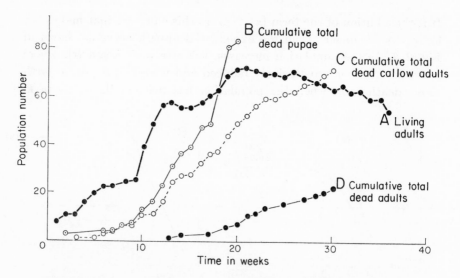

Fig. 2.3 Partial population curves for a culture of *Tribolium castaneum*. Data from Lloyd 1965, culture A5. Lloyd gives separate figures for the predated pupae (B) and newly emerged adults (C) and for the dead fully mature adults (D).

curve was based were therefore 'proved to be true for practical purposes'. When Crombie studied the flour moth *Sitotroga cerealella* under exactly similar circumstances, the moth population did not become stable. In fact we chose to illustrate the logistic curve by Crombie's figures for *Rhizopertha* because they fit much better than any other example we know from the insects. Most observations give a very different picture.

Figure 2.3 shows the results of some studies by Lloyd (1965, 1968) using the flour beetle *Tribolium castaneum*. Here the agreement between the counts of the adult beetles and the best fitting logistic curve is rather poor. But even supposing the fit were excellent, this would not demonstrate that the observations could be explained *only* in relation to the logistic curve. There are in fact a number of formulae which give curves of this general type and Lloyd's detailed observations show how very complex the interactions are between the various stages of the same insect when crowded together. He counted eggs, small and large larvae, prepupae, pupae and adults every week for a large number of replicated cultures. He gave separate figures for dead pupae, dead callow adults (still not fully pigmented) and dead mature adults, which we show as cumulative curves in Fig. 2.3. Figure 2.4 gives for one representative experiment the partial population curves for these stages plotted on a logarithmic scale. These can be compared with the results of the same replicate in Fig. 2.3 where an arithmetic scale was used for the live adults and for the dead adults and dead pupae, which are shown as cumulative curves.

The figures for the adults are fairly constant from the twelfth week onwards, but the numbers of other stages—especially the pupae—show violent changes. Lloyd found that the adult beetles and the larger larvae fed on both the eggs and the pupae. Inspection of the timing of the population changes in Fig. 2.4 shows that whenever the population of large larvae reached a peak, then the number of eggs found was reduced. Note also that the peaks of egg numbers were followed in sequence by peaks in number of small and large larvae, and then by peaks of prepupae and pupae. However, the changes in the population number of pupae have remarkably little effect on adult recruitment. The cumulative number of dead adults in Fig. 2.3D shows that the turnover of mature adults is very slow—many adults produced at the beginning of the experiment must still be alive at the end; their potential longevity is as much as 70 weeks.

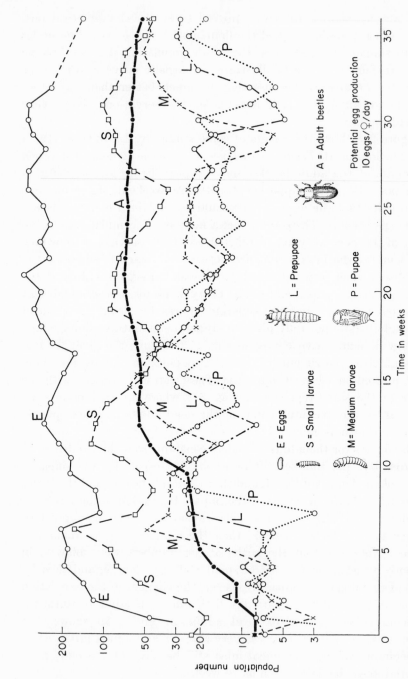

Fig. 2.4 Partial population curves on a logarithmic scale for a culture of *Tribolium castaneum*. Data from Lloyd 1965, culture A5.

This experiment suggests that the attempts to explain the adult curves in terms of the Verhulst–Pearl equation fail partly because the equation completely neglects the changing age structure of the popula-

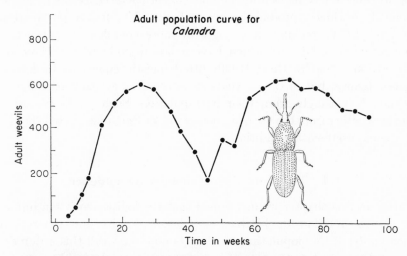

Fig. 2.5 The rice weevil *Calandra* (= *Sitophilus*) *oryzae* was cultured in 12 g of wheat which was renewed every two weeks. Data from Birch 1953.

tion. *N* in equation 2.5 only represents adult numbers. Sang (1950) reached a similar conclusion from a study of populations of fruit-flies.

How then did the simple curve for *Rhizopertha* shown in Fig. 2.1 arise? We suggest as an alternative explanation that the original pair of beetles produced many larvae—perhaps as many as the 10 g of wheat could support. When these in turn become adult, they would emerge over a period of some weeks. If the number of adults emerging per day had a normal frequency distribution and individual adults lived for many months, then the number of adults present at any time during the emergence period would be the cumulative total. A plot of adult numbers against time would be a sigmoid curve very similar in shape to a logistic curve, but its biological implications are quite different. Figure 2.5 shows the adult population curve for the rice weevil *Sitophilus* (= *Calandra*) *oryzae* (Birch 1953). We suggest that the two peaks might merely represent two adult generations (compare Fig. 2.13).

The logistic equation, therefore, seems unsuitable to describe the growth of insect populations where the rate of increase is normally high and the longevity of different age classes is long relative to the time periods considered. Having said this, we should point out that the logistic curve may fit well for populations of bacteria, yeasts or planktonic algae which reproduce by binary fission. To a first approximation it may also fit population changes of more complex animals whose generations overlap and which have a low reproductive increase per generation. Pearl & Reed (1920) fitted logistic equations to human census figures, but in the 50 years since then the actual census figures deviate increasingly from their extrapolation. Such predictive errors can arise either from the use of a poor model, or from unexpected changes in the environmental conditions.

2.4   The definition of density dependence

Earlier in this chapter we mentioned that the logistic equation implies an inverse relationship between population density and the rate of increase (r) of the population. Ecologists nowadays call this a **density dependent** effect on the rate of increase and the ecological literature is full of references to **density dependence** and **density independence,** usually with reference to mortality factors. What do these terms mean?

In their classic study of the gipsy moth and the brown tailed moth, Howard & Fiske (1911) gave the first clear exposition of insect population dynamics. They distinguished three kinds of mortality factor as follows:

1   'A natural balance can only be maintained through the operation of *facultative agencies* which effect the destruction of a greater *proportionate* number of individuals as the insect in question increases in abundance.'

2   'The destruction wrought by storm, low or high temperature, or other weather conditions, is to be classed as *catastrophic*, since they are wholly independent in their activities upon whether the insect which incidentally suffers is rare or abundant.'

3   'Destruction through certain other agencies, notably by birds and other predators are not directly affected by the abundance or scarcity of any single item in their varied menu . . . they average to destroy *a certain gross number* of individuals each year, and . . . work in a manner which is the opposite of "facultative" as here understood' (Howard & Fiske 1911, pp. 107–108, our italics).

The terms we now use for these ideas were suggested by Harry S. Smith (1935). Smith's term *density dependent* is equivalent to facultative and *density independent* is equivalent to catastrophic. Howard and

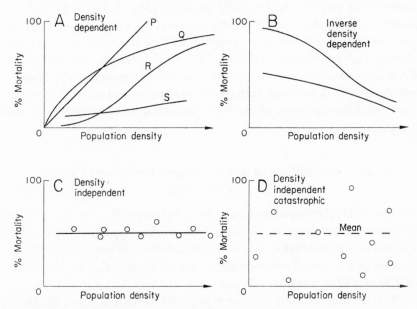

Fig. 2.6 Graphs to describe different relationships between percentage mortality and population density and the terms applied to them.

Fiske's third category, where the proportion killed declines with increasing population density, is now commonly referred to as *inverse density dependent* mortality. These relationships are illustrated graphically in Fig. 2.6. Howard & Fiske (1911), Nicholson (1933) and Smith (1935) and other later authors were right to emphasize the unique position of density dependent factors in stabilizing animal populations. They were mistaken however, in thinking that all relationships which fit the *verbal* definition have the properties needed to regulate populations. (A *regulated* population is one which tends to return to an equilibrium density following any departure from this level.) The situation was further confused because these authors thought that insect parasites (often called parasitoids) act as facultative or density dependent mortality factors. We shall consider them in Chapter 4.

The continuous cultures referred to earlier in this chapter whether they be fruit-flies or beetles feeding in grain or flour are hard to analyse because the generations overlap and many different things are happening at once. In some other insects, especially those which have a single generation each year, the different developmental stages remain separate or overlap rather little in time (Fig. 1.1). It is easier to develop models for such populations and these help us to understand the importance of different mortality factors as causes of population change and population stability. In this chapter we will use simple, discrete-generation population models to demonstrate the ways in which density dependent factors can operate in isolation. Interactions between different types of mortality factors will be considered in later chapters.

2.5   Properties of density dependent factors when acting alone

Ricker (1954) developed a simple way of showing the effects of density dependent mortalities on fishes which he also applied to insects. He constructed 'reproduction curves' in which the egg population $(N_{E_{n+1}})$ of each of a series of generations is plotted against the egg population of the previous generation (N_{En}).

Let us suppose that an insect population with a potential 10-fold rate of increase suffers a single density dependent mortality as defined in Fig. 2.7A. There is no mortality when the initial egg density is below 10, but the k-value for egg mortality rises steeply with increasing egg densities above 10. The equivalent curve of percentage mortality against egg number is shown in Fig. 2.7B. Figure 2.7c shows the corresponding reproduction curve for this population. At low densities the population realizes its potential 10-fold rate of increase, but with populations above 10 the survival is progressively reduced, until when the curve crosses the diagonal line the egg population in successive generations is the same. Clearly, this reproductive curve can be used to read off directly the sizes of successive generations of the insect. If, for example, there are 10 eggs in a given generation we would start at point **a** in Fig. 2.7c by drawing a line parallel to the ordinate, which cuts the reproduction curve at 100. The fate of the next generation is read off starting from **b**. and of further generations from **c, d** and so on. These generations are plotted as histograms in Fig. 2.7D. Had we started with an initial population of 300 eggs (**x** in Fig. 2.7c), the

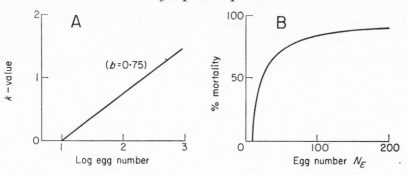

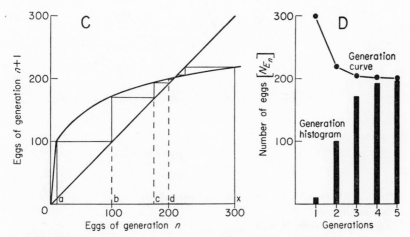

Fig. 2.7
A Defines on logarithmic co-ordinates a density dependent mortality for a population which suffers no mortality when egg density is below 10.
B Shows the equivalent relationship on arithmetical scales.
C Shows the corresponding reproduction curve for this population with a ten-fold increase per generation, from which we can read off the sizes of successive generations of eggs when the initial egg population is represented by point *a* or point *x*.
D Shows a generation histogram for the changes in numbers of the population which starts at *a* and a generation curve for the changes in the population which begins at *x*.

simple construction of a series of steps between the reproduction curve and the diagonal gives the values for successive generations, which in this case decrease towards the stable population of 200 eggs (see

generation curve in 2.7D). The shape of Ricker's reproduction curves
is determined by two things only: the nature of the density dependent
relationship, and the reproductive rate. The important thing about
reproduction curves is that they enable us to see at once whether the

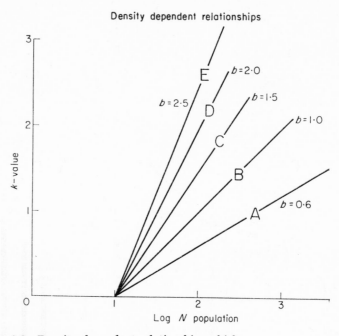

Fig. 2.8 Density dependent relationships which:
Undercompensate A,
exactly compensate B, and
overcompensate C, D and E.

effect of the density dependent factor under consideration will be to
stabilize the population or not.

We find that a new kind of reproduction curve, on logarithmic
co-ordinates, has an advantage over Ricker's curves. The *shape* of the
curve is determined only by the nature of the density dependent factor.
Figure 2.8 shows five density dependent relationships (A–E), each of
which has somewhat different properties. The corresponding logarithmic
reproduction curves for each of them are shown in Fig. 2.9 and may be
used to obtain directly the generation curves shown in Fig. 2.10A–E.
Numerical exercises at the end of the book will help to make these ideas
familiar. These curves have been calculated from the different density

dependent mortalities assuming a reproductive rate (F) of 32 $(\log F = 1\cdot5)$. The models show that different density dependent mortalities can have quite different effects on a population. These

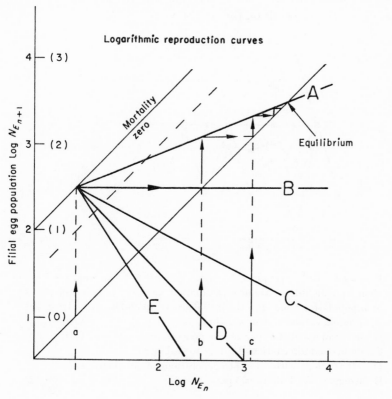

Fig. 2.9 Logarithmic reproduction curves corresponding to curves A–E in Fig. 2.8, calculated by $N_{E_{n+1}} = N_E + 1\cdot5 - k$. Values of k defined by curves A–E in Fig. 2.8.

differences are related to the 'strength' of the density dependent mortality (i.e. the slopes in Fig. 2.8) as follows.

2.5.1 *Undercompensation* $(0 < b < 1)$

The slope b of the line must be between 0 and 1. In Fig. 2.8A the slope is 0·6 and the mortality begins to take effect when the population is greater than 10 (see intercept on X-axis). In the absence of *any* mortality

B

the log population (log N) would increase linearly as shown by the
dotted line in Fig. 2.10 (log $N_{n+1} = \log N_n + \log F$). When an under-
compensating mortality is introduced the population eventually
stabilizes as shown in Fig. 2.10A. The level of this equilibrium or
'steady density' depends on the precise value for the slope and intercept

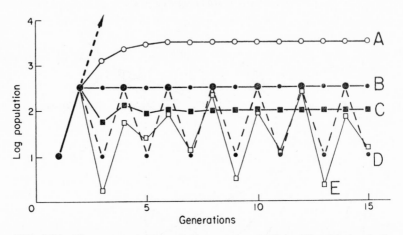

Fig. 2.10 The consequences of the curves A–E in Fig. 2.8 (and of
their equivalent logarithmic reproduction curves A–E shown in Fig. 2.9)
on a population which has the initial size of 10 and in which log F = 1·5.
A   gradual attainment of stability,
B   attainment of stability in one generation,
C   dampened alternations leading to stability,
D   alternations with alternate generations of equal size, and
E   irregular sized alternations.

of the density dependent mortality and on the reproductive rate (F)
of the population. At equilibrium $K = \log F$. The time taken to reach this
level also depends on these as well as on the starting point for the model.

2.5.2   *Exact compensation* ($b = 1$)

The density dependent relation must have a slope of $b = 1$ (Fig. 2.8B).
Within limits such a mortality perfectly compensates for any population
change. Thus, if the disturbing factor or factors act prior to such a
density dependent factor, the population in each generation will
return to the same level (Fig. 2.10B). Recovery from very big reductions
will take more than one generation.

2.5.3 *Overcompensation* $(b > 1)$

If the density dependent slope is between 1 and 2 ($b = 1.5$ in Fig. 2.8c) the population tends toward an equilibrium value but initially the successive generations alternate with decreasing amplitude above and below that value (Fig. 2.10c).

With a density dependent slope of 2 (Fig. 2.8d) any deviations from the population equilibrium will result in alternations between the same high and low values as shown in Fig. 2.10d.

From Fig. 2.10a to d one might expect a density dependent mortality with a slope greater than 2 to cause alternations of increasing amplitude in the population following any disturbance from the equilibrium level. However, the situation is complicated because the population soon falls below the level at which the mortality acts. In Fig. 2.10e this occurs in generations 3, 9, and 13 so that in generations 4, 10 and 14 the populations have increased 32-fold ($F = 32$). Having a *threshold density* below which there is no mortality prevents the population becoming entirely unstable—it fluctuates violently but within limits.

The models in Fig. 2.10 serve to illustrate that within definable limits density dependent factors tend to stabilize a population; outside these limits they may over-react so strongly to population change that they can be the cause of considerable population fluctuation. In practice, a density dependent factor may be too feeble to stabilize a natural population which is affected by a strongly variable density independent mortality.

The interactions between different kinds of mortality, which we study in later chapters, can be understood only if life tables are detailed enough to separate out the different kinds of mortality. Reproduction curves are of limited application because they do not make this separation, and are concerned only with the total mortality within the generation.

2.6 Contest and Scramble competition

The effect of intraspecific competition is always measurable as a density dependent process. For example, competition for food may result in density dependent mortality, reduction in fertility or dispersal.

A.J.Nicholson (1954) recognized two extreme forms of competition which he called 'contest' and 'scramble' competition.

In *contest* each successful animal gets all it requires—the unsuccessful animals get insufficient for survival or reproduction. An often quoted theoretical example is the competition between solitary wasps for a limited number of nest holes. Figure 2.11A shows the expected result of a contest for 100 nest holes. With 100 or fewer competitors, there is no shortage. With 200 competitors only 100 can find nest holes and therefore 50 per cent fail to breed (k-value = 0·3). If there are 1000 competitors, still only 100 succeed in finding nest holes and now 90 per

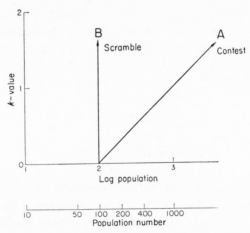

Fig. 2.11 The mortality relationships associated with two extreme kinds of competition—scramble and contest.

cent fail to breed (k-value = 1·0). It is characteristic of this idealized contest competition that the k-value rises with a slope of unity when compared with the log population densities above the level at which the resource begins to be in short supply. In Chapter 3 we shall find examples in which the k-value rises less steeply than this—further research will probably show a far wider range of effects than suggested by Nicholson's simple classification.

'The characteristic of *scramble* is that success is commonly incomplete, so that some, and at times all, of the requisite secured by the competing animals takes no part in sustaining the population, being dissipated by individuals which obtain insufficient for survival' (Nicholson 1954, p. 42).

The resource is therefore being shared amongst all the competing animals. With identical animals in a uniform habitat the sharing would be equal and the mortality would rise immediately from 0 to 100 per

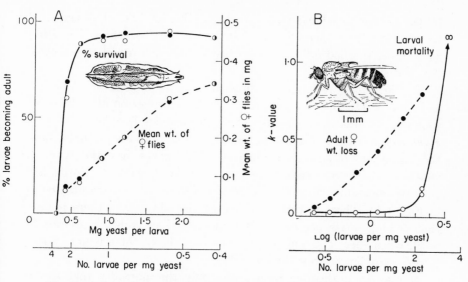

Fig. 2.12 Competition for food in the fruit-fly *Drosophila*. The figures from Bakker (1961) are plotted:
A to show the effect of food per larva on survival and adult weight,
B to show how larval mortality and adult weight loss, both expressed as *k*-values, are related to the logarithm of the number of larvae per mg of yeast.
● Constant food, 120 mg yeast; number of larvae varied.
○ Constant larvae, 100; amount of food varied.

cent when the resource (food for example) per individual becomes just insufficient for survival. The mortality in this idealized situation would take the form shown in Fig. 2.11B. If in reality there were some members of the population better able to secure food than others, the actual curve would rise less abruptly than in Fig. 2.11B.

We shall now consider two examples of competition in insects which illustrate these ideas.

Bakker (1961) made a careful experimental study of competition between larvae of the fruit-fly *Drosophila melanogaster*. Known numbers of newly hatched larvae were put on to an agar surface with known amounts of yeast on which the larvae fed. Varying the numbers of larvae on a fixed quantity of food, or varying the amount of food for a fixed number of larvae, gave the same result—it was the food supply per larva that was critical. When survival is plotted against the weight of food per larva the survival rose from zero with 0·3 mg food to about

90 per cent with 0·6 mg yeast per larva (Fig. 2.12A); but the adult female flies which emerged from these larvae were half starved and very small. Flies reached full size only when larvae each received more than 2 mg yeast.

In Fig. 2.12B we have re-expressed Bakker's data in the same way as in Fig. 2.11. The larval survival has been converted into the corresponding k-values and the weight of the females is expressed as the k-value equivalent to the loss in female weight. Both are plotted against the log density of larvae per mg yeast. The curves show the extent to which mortality and weight loss are density dependent. Notice also that the mortality curve in Fig. 2.12B resembles the scramble curve in Fig. 2.11. There is very little larval mortality when there is more than 0·6 mg yeast per larva. With less food mortality rises rapidly until, when there is 0·3 mg or less yeast per larva there is not enough food for any larvae to survive.

It may be misleading to measure only the mortality that results from competition. Figures 2.12A and 2.12B show clearly that adult female weight, and therefore probably the egg production per female, depends on the food available to the larvae. Between 0·6 and about 2 mg yeast per larva is sufficient for survival, but the adults are sub-optimal in size. Many insects seem to be able to compensate for some food shortage in this way.

For our second example we have re-examined some observations published by Nicholson (1954) on laboratory populations of the Australian sheep blow-fly, *Lucilia cuprina*. This insect is almost indistinguishable from the common green-bottle flies whose larvae attack either living sheep or mammal corpses in many parts of the world including Europe and the USA.

Nicholson kept the flies in large cages in the population experiment under consideration and the adult flies were provided with excess of food (ground liver and sugar) so that they laid plenty of eggs. But the food for the larvae was restricted by adding to the cage each day only a small pot containing 50 g of meat. The flies could lay their eggs on this meat but were unable to feed on it. At a suitable time each pot was examined and the number of viable puparia was counted; adults were allowed to emerge from the puparia and were then added to the cage. The results of counting adult flies every two days are shown in Fig. 2.13. The fly population passes through seven well marked peaks. Nicholson termed them 'oscillations'.

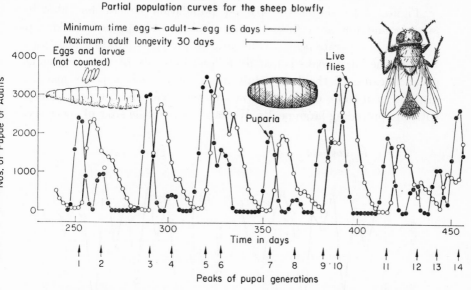

Fig. 2.13 Partial population curves for puparia and adults of *Lucilia cuprina*. Larval food supply was 50 g meat per day. Ground liver and sugar were supplied in excess of needs for the adult flies. Data from from Nicholson (1954).

Nicholson (1933) had put forward a theory which predicted population oscillations in parasitic insects and their hosts; this is discussed in detail in Chapter 4. Nicholson's use of the term 'oscillation' both for the theoretical parasitic-host population changes and for these changes in blow-fly population implies a close similarity in the form of the population changes. We shall now show that these types of population change are quite unrelated and require different types of explanation.

The first basic difference is that in Nicholson's theoretical work on parasites, the oscillations appear when the figures are plotted as an *adult generation curve*. The peaks are five or more generations apart. Nicholson's blow-fly 'oscillations' were plotted as an *adult population curve* (Fig. 2.13). We saw in Chapter 1 that the same information plotted in these two different ways looks entirely different. Conversely, any similarity between the figures which have been plotted in different ways is likely to disappear when the two are plotted using the same conventions. To make a proper comparison we must replot the blow-fly data as a generation curve.

Figure 2.13 shows the partial population curves for blow-fly puparia and for adult flies. In the numbers of puparia it is easy to recognize 14 separate peaks, which must represent separate generations. Each of the large peaks is usually followed quickly by a much smaller peak. Because of the long adult life span the adult peaks are less distinct than those of the puparia and the seven smaller peaks merge with the preceding large peaks. We can estimate the sizes of successive

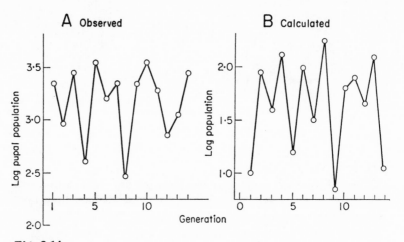

Fig. 2.14
A   Observed logarithmic generation curves for sheep blow-fly puparia where 50 g food is supplied per day to the larvae. Data in Fig. 2.13.
B   Model calculated for 1 g of food per generation. Data in Fig. 2.15A.

generations more easily from the data for the puparia than for the adult flies. Southwood (1966) describes some methods of estimating the numbers entering any developmental stage, but to use them we require some biological data not given by Nicholson. We have therefore made a rough estimate of the size of each generation from the peak number of puparia, which can be read off Fig. 2.13. They are plotted as a generation curve in Fig. 2.14A. The fluctuations in the generation curve are irregular with peaks now at intervals of 2, 2, 2, 3, and 4 generations. These changes are entirely different from the regular oscillations in numbers in a parasitic-host interaction, where population peaks are usually more than five generations apart (Chapter 4).

In the same paper Nicholson (1954) described a simple competition experiment in which known numbers of newly hatched blow-fly larvae were put with 1 g of meat to serve as food. The outcome shown in Fig. 2.15A was a maximum production of about 16 flies when 30 larvae competed for the food. When there were 80 larvae originally present the average of between 5 and 6 became adults, and there were no survivors when 200 or more larvae competed. This is clearly an example of scramble competition: when the figures are expressed as changes in the k-value with increasing initial population, the slope of the curve rises very steeply (Fig. 2.15B) and can be compared with Fig. 2.11. Earlier

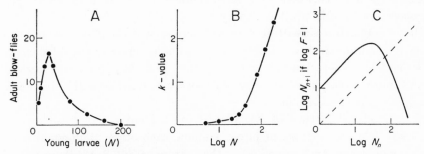

Fig. 2.15 The effect of competition between larvae of the sheep blow-fly for 1 g of food is plotted in three ways:
A as graphed by Nicholson (1954),
B k-value plotted against log N shows overcompensation,
C the logarithmic reproduction curve for 10-fold increase.

in this chapter we showed that overcompensating density dependent mortalities can produce alternations in numbers in successive generations. The consequences of this particular density dependent relationship can best be investigated for a population by constructing the corresponding logarithmic reproduction curve. This is scaled for 10-fold increase per generation in Fig. 2.15c. Clearly it would cause the population to be unstable; from Fig. 2.15c we can read off population densities for successive generations using the construction introduced in Fig. 2.7c, and one set has been plotted in Fig. 2.14B for comparison with the observed values for a diet 50 times as great. The level is very close to what might be expected, and the amount of change is very similar, as each has peaks at irregular intervals of 2, 3 or 4 generations. Although this comparison does not provide a very critical test it indicates that fluctuations of the kind observed in these blow-fly experiments could result from the very strongly density dependent mortality caused by larval competition. The fluctuations are entirely unlike the parasite induced oscillations considered in Nicholson's earlier work (1933) which we shall discuss in Chapter 4.

The graphical interpretations based on Nicholson's very simple experiment enable us to predict the outcome of his much longer and more costly experiments. In one of these the adult flies were subjected to 99 per cent mortality on emergence. We cannot predict the effect of this from Fig. 2.15c very accurately, because we do not know how such imposed mortality affects the mean reproductive rate of the surviving flies. However, if F were reduced to a two-fold increase per generation by this adult mortality, we can see by rescaling Fig. 2.15c that a diagonal representing two-fold increase would cut the curve near its maximum, which would indicate that the population should then be stable from generation to generation. The observed populations were in fact more stable, but the main difference was that they did not break into separate generations.

One minor feature of Nicholson's experiment remains to be discussed. The generations of puparia follow each other at unequal time intervals, which are marked off at the bottom of Fig. 2.13. Because the daily ration of larval food can produce viable maggots which develop into puparia only when eggs are few, puparia will appear about 10 days after any period when flies have been few—either when a new emergence has begun (day 250) or when two combined adult generations eventually die of old age. Alternate generations effectively reproduce either when the flies are young, or when at the very end of their reproductive lives, so that the interval between peaks of puparia alternates between about 12 days and 22 days. Towards the end of this experiment and of other similar experiments described by Nicholson, the flies seem to behave rather differently from at the beginning. Probably the extremely severe larval competition has selected out some rare genetic combination which is normally at a disadvantage.

CHAPTER 3

COMPETITION BETWEEN SPECIES FOR A
LIMITED RESOURCE

3.1 Synopsis

When two species compete for the same resource in a culture usually only one survives and the other becomes extinct, but on occasion co-existence of both species has been observed. The competition equations of Lotka and Volterra, developed from the simple logistic model for intraspecific competition, may predict either co-existence, or the elimination of one species or the other depending precisely on how the species affect each other.

Co-existence has been observed in experiments where the Lotka–Volterra model predicts the elimination of one of the species. This is because the assumption that the effects of intraspecific and interspecific competition are proportional to population density is not satisfied. Such non-linear density dependent relationships can be demonstrated in experiments.

In the field, closely related species which use the same food resource often live together and non-linear density dependent effects of this kind may provide the mechanism for this species diversity. However, the adverse effects of competition in the field have rarely been quantified and it has not been easy to assess the contribution to stability provided by the small observed ecological differences between the species which live together. Non-linear density dependent mortality may also be important in explaining some genetic polymorphisms in natural populations, since the same mortality may be involved both in the selection of genetic alleles and in population regulation.

The process of competitive exclusion has been observed with certainty under field conditions only when several species of

33

parasitic hymenoptera have been introduced successively into a region, where they competed for the same host species.

3.2 Introduction

When two or more different species compete for limited resources such as food or space, interspecific competition will be superimposed upon the intraspecific competition which we examined in the previous chapter. Here we shall consider only laboratory experiments because we cannot hope to interpret correctly the results of field studies, where closely allied species are frequently found to live together, until we understand the relatively simple interactions between two species under constant conditions.

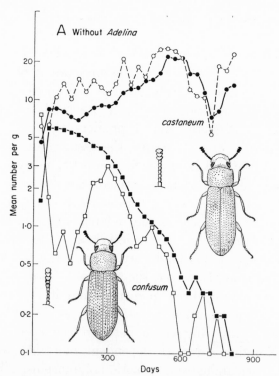

Fig. 3.1 Competition between two species of *Tribolium* at 29·5°C on wholemeal flour with 5 per cent yeast.

A *Tribolium castaneum* eliminates *T. confusum* in the absence of *Adelina*. Mean of 12 replicates in 8 g flour. Data from Park (1948), Table 27 I EaS.

3.3   Experiments with competing species

Extensive studies of competition between two species of flour beetles, *Tribolium castaneum* and *Tribolium confusum* were conducted by Park (1948). He started his mixed cultures in sifted flour with known numbers of adult beetles and counted the adults, large larvae and pupae every 30 days. When *Tribolium castaneum* and *confusum* were mixed together, one of the two became extinct after about a year and the other species assumed the population density it would normally have reached alone (Fig. 3.1A).

Park (1954) found that the experimental conditions were very important in determining which of the two species would replace the other. At temperatures above 29°C *T. castaneum* was favoured, but below 29°C *T. confusum* was usually the successful species. The results were also considerably affected by the presence or absence of a parasite

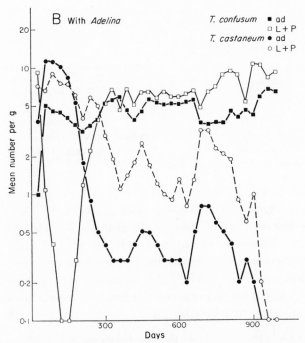

B *Tribolium confusum* eliminates *T. castaneum* in the presence of *Adelina*. Mean of 9 replicates in 40 g flour. Data from Park (1948), Table 16 II Ea.

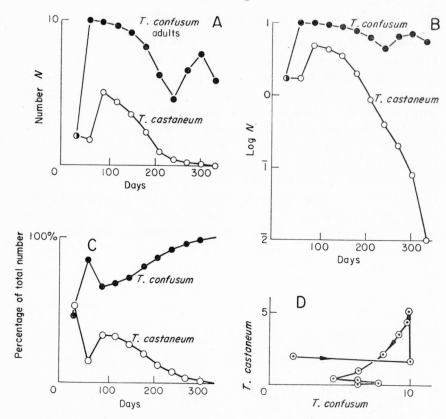

Fig. 3.2 Four ways of graphing the competition between two species of *Tribolium* in 80 g of wholemeal flour and 5 per cent yeast. Data from Park (1948), Table 20, replicate III Eb.

of the beetles, a protozoan named *Adelina**, to which *T. castaneum* was particularly susceptible. Figure 3.1A shows how at a constant temperature of 29·5°C and in the absence of *Adelina*, *T. castaneum* tended to eliminate *T. confusum*. However, in infected cultures at the same temperature, *T. castaneum* was eliminated by *T. confusum* in about nine out of 10 replicates as shown in Fig. 3.1B. In those few replicates in which *T. castaneum* was the surviving species, it reached populations only a third as dense as when uninfected.

* The protozoan *Adelina* is an intracellular parasite belonging to the Microsporidia; it multiplies especially in the mid-gut cells of the host. The spores pass out of the host with the faeces and can then be ingested by other larvae which thus become infected.

In Fig. 3.2 we show the results for another replicate (in which *Adelina* was present) plotted in four different ways, each of which emphasizes different features of the interaction. In Fig. 3.2A the population scale is arithmetical. In Fig. 3.2B the population is plotted on a logarithmic scale. Now we see clearly that the *rate* of the decrease in the *T. castaneum* population is steadily increasing. In Fig. 3.2C

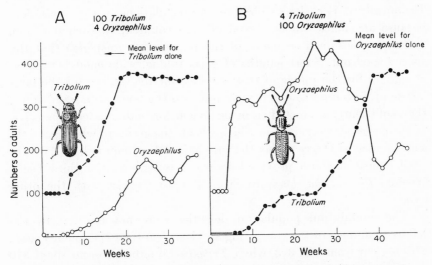

Fig. 3.3 Competitive co-existence of two kinds of beetle in a renewed quantity of wheat. From the different initial conditions shown in A and B the end result is the same with some 370 *Tribolium* and 180 *Oryzaephilus* co-existing. Data from Crombie (1946).

the percentage of each species in the population is plotted against time. One curve is then the inverse of the other. Lastly, in Fig. 3.2D the competition process is followed in an interesting way by plotting the numbers of one species on the abscissa and the other on the ordinate. We shall see (Figs. 3.4, 3.5) that this last method, which eliminates time from the graph, gives us a new theoretical insight into the processes involved.

Gause, in his book *The Struggle for Existence* (1934) showed by experiment that two species of competing Protozoa (*Paramecium aurelia* and *P. caudatum*) did not co-exist if they relied on the same resources. One species eliminated the other; this has become known as *competitive exclusion* or as 'Gause's Principle' 'Two species competing for limited resources can only co-exist if they inhibit the growth of competing species less than their own growth' (Ayala 1970).

Crombie (1946) described the growth of populations of two kinds of beetle, *Oryzaephilus surinamensis* (the saw-toothed grain beetle) and *Tribolium confusum* (the flour beetle). The food medium used was either wheat grains, or wholemeal flour that had been coarsely or finely ground. In all cases the medium was renewed at suitable intervals. Technically it is easier to observe the populations in the finest flour because all stages can be sieved off and counted. In such fine flour *Tribolium* consumed so many of the pupae of *Oryzaephilus* that this species died out. Short lengths of glass tubing, wide enough to serve as refuges for the pupae of *Oryzaephilus* but too narrow for the large larvae of *Tribolium* to enter, were put into the flour. *Oryzaephilus* was then sufficiently protected from predation for both species to survive. Both species also survived in whole wheat, the grains of which gave the small pupae of *Oryzaephilus* the protection they required. In fact the whole-wheat medium provided so many refuges that a third beetle species—*Rhizopertha dominica*—was able to co-exist with the other two.

The equilibrium population densities were not affected by the relative numbers of the competing species when the experiment began. This is clear from Fig. 3.3, where *Tribolium* numbers rose to about 370 and *Oryzaephilus* to 180 whichever species had initially been given the advantage of numerical preponderance. Crombie's experiments show that in the presence of suitable refuges, two or more species which share the same food can co-exist. The beetles concerned belonged to different families, but there is no reason to believe that such taxonomic separation is necessary. Under field conditions closely allied and indeed congeneric species can co-exist and Ayala (1969, 1970) has recently shown that competing species of *Drosophila* can co-exist under carefully controlled conditions. He employed a routine subculturing method in which adult flies were periodically removed, identified, counted and put on fresh medium, whilst the old medium, including larvae and pupae, was retained until more adults emerged, which were likewise counted and added to those which had emerged previously. Ayala performed these experiments at a number of different temperatures and (like Park) he found that competitive exlusion was the general rule. At and below 19°C *Drosophila pseudo-obscura* thrived and replaced *D. serrata* but above 25°C *D. serrata* eliminated *D. pseudo-obscura*.

Park had found that at an intermediate temperature it seemed largely a matter of chance which species of *Tribolium* was successful. Ayala found that at 23·5°C both species were able to co-exist indefinitely when subcultured in the routine way.

To generalize the situation let us call the competing species N_1 and N_2. As Crombie and Ayala have shown, under some conditions N_1 eliminates N_2; under other conditions N_2 eliminates N_1. Between these conditions it may either be a matter of chance which survives, or there may be a zone of conditions in which the two species can co-exist.

3.4 The basic theory of competition

Lotka (1925, 1931, 1932) and Volterra (1931) were able to define in precise mathematical terms the conditions for results of these kinds by extending the equations of Verhulst and Pearl to cover two competing species. Designating the competing species N_1 and N_2 we can add a new term to equation (2.5) to describe the idea that the relative rate of increase $(1/N)$ (dN/dt) has a maximum value of r_m, but is reduced by some unspecified function of both the population of N_1 and the population of N_2.

$$\frac{1}{N_1}\frac{dN_1}{dt} = r_{m1} - f(N_1) - f(N_2)$$

$$\frac{1}{N_2}\frac{dN_2}{dt} = r_{m2} - f(N_2) - f(N_1) \tag{3.1}$$

For mathematical convenience, but without support of biological measurement, it has been customary to consider the case where the various unspecified functions are all simple proportions. Thus, when κ_1 and κ_2 represent the equilibrium populations of each species alone, we can write

$$\frac{1}{N_1}\frac{dN_1}{dt} = r_{m1}\frac{(\kappa_1 - N_1 - \alpha N_2)}{\kappa_1}$$

$$\frac{1}{N_2}\frac{dN_2}{dt} = r_{m2}\frac{(\kappa_2 - N_2 - \beta N_1)}{\kappa_2} \tag{3.2}$$

where α and β are constants ('competition coefficients'). Gause & Witt (1935) rightly say that these equations are true only for very simple populations like those of yeast cells. However, in spite of the fact that the competitive relations may not be expressed accurately by equations like (3.2), these equations are much quoted because they help us to picture the way in which competition may act.

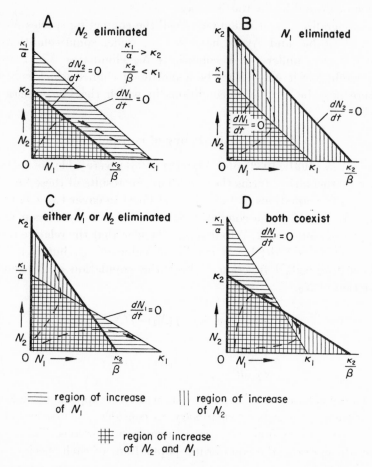

Fig. 3.4 The four consequences of the Lotka–Volterra equations.

Lotka (1932) distinguished four possible consequences of these equations, which were first shown diagrammatically by Gause & Witt (1935) and are illustrated in Fig. 3.4. With N_2 as the ordinate and N_1 the abscissa the lines representing the loci of $dN_1/dt = 0$ (i.e. where the population of N_1 does not change) and $dN_2/dt = 0$ are plotted by finding the values for intercepts as shown in the figure (κ_1, κ_2, κ_1/α, and κ_2/β). Note that the steepness of these lines is a measure of the intraspecific effects of competition; the steeper the line, the more inhibition within the species.

The outcome of competition can now be predicted from the relative positions of these two lines. When the line for species N_1 lies 'above' that for N_2 as in Fig. 3.4A, species N_1 will eliminate N_2. For example, starting with small numbers of a mixed population, both species will initially rise in numbers until the co-ordinate (N_2, N_1) crosses the line representing $dN_2/dt = 0$. In the zone bounded by the two lines, species N_2 will decrease and N_1 will increase until the point when N_1 reaches its maximum population size κ_1; N_2 now becomes extinct. The situation in Fig. 3.4B is exactly the reverse: N_2 now eliminates species N_1. In Fig. 3.4c the two lines cross each other at a point which is an equilibrium position where neither species changes in numbers. However, this is an unstable equilibrium because each species inhibits the other more than itself and the least deviation will eventually lead to the extinction of one or other of the species. The outcome of competition in such cases would normally depend on the relative size of the initial populations of the two species. In Fig. 3.4D we have the reverse to that in Fig. 3.4c: a stable equilibrium where the lines cross and the population trends always converge on this point. This is because each species inhibits its own increase more than that of the other species.

Crombie (1945, 1946) showed the extent to which Lotka and Volterra's equations predicted the outcome of several experiments on the competition between stored product insects. Figure 3.5 shows the observed outcome of competition between two grain beetles, *Tribolium* (N_1) and *Oryzaephilus* (N_2). Crombie was able to measure the equilibrium population densities of each species alone, (κ_1 and κ_2) and calculated the competition coefficients (α and β) from which the loci of $dN_1/dt = 0$ and of $dN_2/dt = 0$ were determined. The relative position of these loci correspond to that in Fig. 3.4D, so we expect both species to co-exist in a stable equilibrium, at the point where the lines cross. Figure 3.5 shows that co-existence did occur in Crombie's experiments at approximately the calculated levels and irrespective of the initial densities of N_1 and N_2.

3.5 Results which conflict with theory

When Ayala (1969, 1970) studied two species of *Drosophila*, *D. pseudo-obscura* and *D. serrata*, he was able, like Crombie to estimate from his

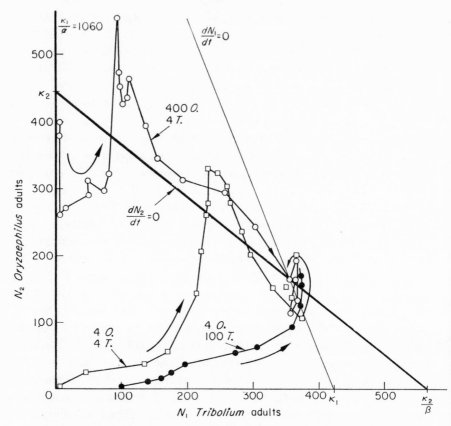

Fig. 3.5 The observed outcome of competition between *Tribolium* (*T*) and *Oryzaephilus* (*O*) is the same for three combinations of initial densities of the two species. The populations co-exist and the graphs resemble Fig. 3.4D. Data from Crombie (1945, 1946).

data the necessary constants in equation (3.2), but unlike Crombie, he found that the theory failed to explain the co-existence of the two *Drosophila* species.

Ayala's results (Fig. 3.6) show that $\kappa_1/\alpha < \kappa_2$ and $\kappa_2/\beta < \kappa_1$. These are the conditions in Fig. 3.4c where it is chance which species survives, but both cannot do so! There are various unreal assumptions in these mathematical equations which may account for this discrepancy. For instance, 'age structure' in the population is neglected (see Chapter 2). More important still, however, is the tacit assumption of linearity in the functional relationships in equations (3.2). The jump from the

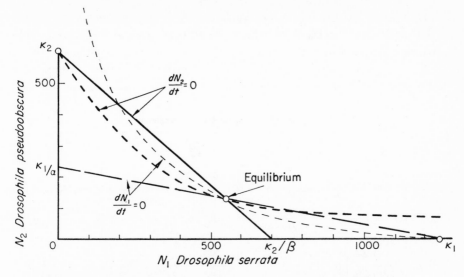

Fig. 3.6 Conditions for the co-existence of two species of *Drosophila*. Data from Ayala (1969, 1970). Straight lines, calculated loci; broken curves, assumed real loci.

general relationship in equation (3.1) to the linearity assumed in the αN_1 and βN_2 terms representing competition could well be biologically wrong, even if mathematically convenient. Ayala's method of fixing the loci $dN_1/dt = 0$ and $dN_2/dt = 0$ assumed that these loci are straight lines. His experimental results provided values for κ_1 and κ_2 and the size of both populations at equilibrium. The calculated points (κ_1/α) and (κ_2/β) were in effect obtained by extrapolating the lines between κ_1 and κ_2 and the equilibrium point to the respective axes as shown in Fig. 3.6 (solid lines). Gilpin & Justice (1972) have recently confirmed our view that the real lines representing the loci of $dN/dt = 0$ must be curved as shown in Fig. 3.6.

How can we test the relations assumed in the Lotka–Volterra equations? For this purpose equations (3.1) can be rewritten in a simplified form for species N_1.

$$\frac{1}{N_1}\frac{dN_1}{dt} = r_{m1} - \alpha N_1 - \beta N_2 \qquad (3.3)$$

where α and β are constants.

It is hard to measure instantaneous rates of increase (dN/dt) since a finite time interval is needed to get a measurable change. If instead we take a fairly long time, such as one complete generation, and consider the special case where generations are synchronized, conditions are not in every way the same as assumed in equation (3.3), because some possible interactions are avoided by the changing age structure of the population. However, if there is a linear relation

Effect of larval crowding on adult emergence

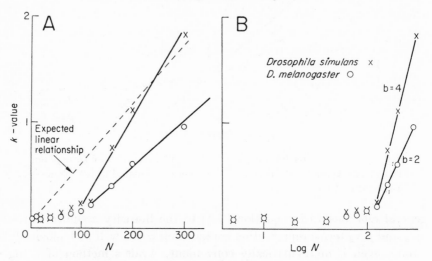

Fig. 3.7 Different numbers of larvae of either *Drosophila simulans* or *D. melanogaster* were placed on a fixed quantity of food and the numbers of emerging adults were counted. The k-value for the resulting mortality is plotted in A against the initial density of larvae and in B against the log initial larval density. B shows that above a population of 100 both species suffer from an overcompensating density dependent mortality. Data from Miller (1964).

between the rate of increase in equation (3.3) and N_1 and N_2, we can argue by analogy that the case with synchronized generations should be represented by

$$\frac{N_{1n+1}}{N_{1n}} = F - aN_{1n} - bN_{2n} \qquad (3.4)$$

where N_n and N_{n+1} are the numbers of adults of species N_1 in the parental and filial generations of N_1, F is the maximum production of offspring per insect, and aN_{1n} and bN_{2n} represent the mortality caused by intraspecific competition and interspecific competition respectively. Clearly aN_n represents a density-dependent mortality of the form of curve P shown in Fig. 2.6A. Similarly the mortality of N_1 caused by N_2 should be linearly related to the population of N_2.

We saw in the last chapter that measurements for the effects of intraspecific competition in *Drosophila* were linear over a certain range on a log–log scale (Fig. 2.12B) which implies a curvilinear relation if plotted as percentage mortality against population number. This is confirmed for larval survival in *Drosophila* by Miller (1964) who put different numbers of larvae of either *Drosophila simulans* or *D. melanogaster* into a fixed quantity of food and scored the numbers of adults which emerged. In Fig. 3.7A the mortality in *Drosophila* between the larval and the adult stage is plotted as a k-value against the initial number of larvae on an arithmetic scale. Mortality is low until 100 larvae are used and beyond this threshold the k-value rises roughly in proportion to the initial larval density. The rise is much steeper for *D. simulans* than for *D. melanogaster*. (In Fig. 3.7B the k-value is plotted against the logarithm of the population. Above a population of 100 the rise in k-value has a slope which strongly overcompensates [compare Fig. 2.8]). The dotted line in Fig. 3.7A shows the expected form for the curve if the rate of increase $dN/dt = rN - \alpha N^2$.

Birch, Park & Frank (1951) have shown for *Tribolium* how net egg production is affected by adult numbers. The data do not conform to the Lotka–Volterra equation. Cultures with equal numbers of males and females were set up and the number of eggs present each week were counted in 8 g flour for eight weeks. Eggs and larvae were removed. Table 3.1 gives the net number of eggs per beetle per day for two species of *Tribolium* in isolation. We expect from equation (3.4) that

Table 3.1 Birch, Park & Frank (1951) counted the eggs laid per day by *Tribolium confusum* and *T. castaneum* at different densities over a period of eight weeks. We have taken means from their Table 1.

Number of beetles	2	16	80	160
T. confusum eggs/beetle/day	5·4	3·2	1·2	0·75
reduction expressed as a k-value	—	0·23	0·65	0·86
T. castaneum eggs/beetle/day	7·5	4·0	1·5	0·87
reduction expressed as a k-value	—	0·27	0·70	0·93

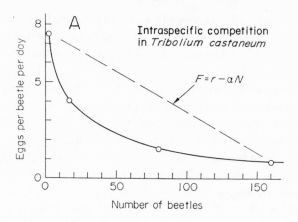

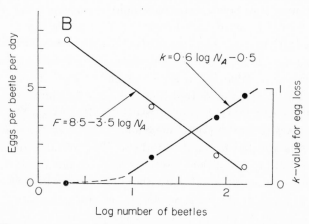

Fig. 3.8

A In *Tribolium castaneum* the reduction in number of eggs laid per beetle per day is not proportional to the density of adult beetles.

B Daily egg production per beetle is fitted well by a straight line when plotted against the logarithm of the number of beetles. The k-value for egg loss is not proportional to the logarithm of beetle numbers, but three of the four points fit a straight line. Data from Table 3.1.

the eggs per beetle will decline in linear proportion to beetle population. This was not the case. The decline in eggs per beetle F is clearly curvilinear when plotted against beetle population density N_A as shown in Fig. 3.8A. When plotted against the logarithm of the population, the points are fitted by the equation

$$F = a - b \log N_A \tag{3.5}$$

When plotted as a k-value for egg loss against the log of the number of beetles, three of the points fit the relation

$$k = 0\cdot6 \log N_A - 0\cdot5 \qquad\qquad (3.6)$$

Curves for *T. confusum* are similar. These are entirely different relationships from that assumed by Lotka and Volterra.

Mertz & Davies (1968) studied the destruction of pupae of *Tribolium castaneum* by the adults. Neither the numbers killed nor the k-values we have calculated for the pupal mortality show simple relations with beetle numbers of the kind that the Lotka–Volterra equations postulate. At low pupal density, almost all were destroyed by the beetles; but when large numbers were added many survived because the appetites of the adults were satisfied—here we have evidence of inverse density dependence at high pupal numbers, which is another kind of deviation from the simple linearity assumed by Lotka and Volterra.

These tests for the linearity of the effect of population density on rates of increase have all been for intraspecific competition and none has covered the effect over a whole generation. As a class exercise in Oxford we set up competition experiments which can be analysed to

Table 3.2 Results from a class experiment with two kinds of beetle competing in a small quantity of flour enriched with yeast. Experiments start with equal numbers of males and females.

Initial number of *Cryptolestes* N_1		4	8	16	32	64	128
A	Number of adult progeny	101	180	276	427	411	473
B	*Cryptolestes* with 16 *Cathartus*, N_2						
B_1	Number *Cryptolestes* adult progeny	86	260	208	414	430	
B_2	Adult progeny of the 16 *Cathartus*	208	185	164	121	78	
Initial number of *Cathartus* N_2		4	8	16	32	64	128
C	Number of adult progeny	96	287	331	333	434	392
D	*Cathartus* with 16 *Cryptolestes*, N_1						
D_1	Number of *Cathartus* adult progeny		74	182	192	275	
D_2	Adult progeny of the 16 *Cryptolestes*	344	306	199	356	228	

measure both intraspecific and interspecific effects. We used two kinds
of beetle, *Cathartus* and *Cryptolestes*, which we chose because of their
rapid development—adults produce adults of a new generation in

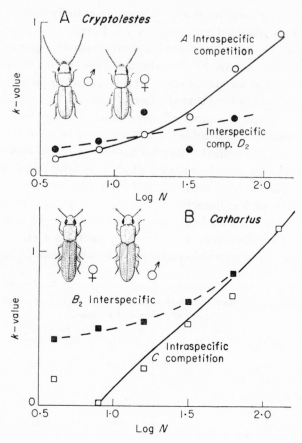

Fig. 3.9 Intra- and interspecific competition between species of
Cryptolestes and *Cathartus*. Data from Table 3.2.
A *k*-values for *Cryptolestes*,
B for *Cathartus*.

about six weeks at 25°C. The sexes are easily recognized by the antennae
or by the shape of the thorax. We measured out small amounts of flour,
enriched with yeast, into glass vials and introduced 4, 8, 16, 64 and
128 beetles. In experiment A all were *Cryptolestes*. In B, 16 *Cathartus*
were present in addition to the *Cryptolestes*. In experiments C and D

the species were reversed. The numbers of both species were counted when the filial generation had become adult, and in Table 3.2 the numbers of the filial generation are recorded. These figures can be analysed in various ways: As a test of linearity, we can select from Table 3.2 figures for the rate of increase of *Cryptolestes* in relation to *Cryptolestes* numbers when *Cathartus* is either O(A) or constant at 16 (B$_1$). The effect of *Cathartus* on *Cryptolestes* is shown when the initial numbers of *Cryptolestes* are kept constant and the numbers of *Cathartus* are varied (D$_2$). These are plotted in Fig. 3.9A; Fig. 3.9B shows similar plots for *Cathartus* alone and with different numbers of *Cryptolestes*.

If equation (3.4) were valid, then in Fig. 3.9A, B, each set of points should be fitted by a straight line. There is evidence for heterogeneity in the figures, especially where both species were concerned, but the general trend of the points suggests that linearity would be improved by a change of scale. If the information is replotted with N_{n+1}/N_n on the ordinate and $\log N_n$ instead of N_n on the abscissa the points for each species alone are approximately fitted by the line

$$\frac{N_{n+1}}{N_n} = 36 - 15 \log N_n \qquad (3.7)$$

which is shown in Fig. 3.10.

This kind of relation is quite different from that postulated in the Lotka–Volterra equations. It is exactly the right kind to give the curved loci for $dN/dt = 0$ that we had to assume for *Drosophila* in Fig. 3.6. This explains why the equilibrium position for the mixed population of the two *Drosophila* species was so much lower than either population could maintain alone.

3.6   Interspecific competition in the field

Field observations will normally show only the final result of competition. The process of exclusion may be far back in the past and direct evidence for it comes only from a study of introduced species. Closely allied species which exploit the same resources often seem to co-exist in a stable way. In England there are five common species of social wasp of the genus *Vespula*, which are all of the same size and must compete for food; how they co-exist is not fully understood. There are many species of caterpillar which feed on young oak leaves in spring. They must compete for food at least in those years when the trees are defoliated. We think that they co-exist because each is regulated by specific parasites at such a low density that collectively they seldom run short of food.

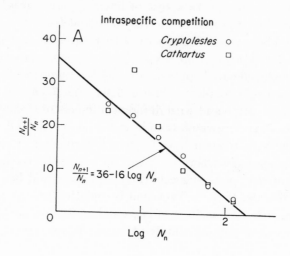

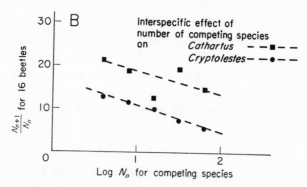

Fig. 3.10 The information in Fig. 3.9 is replotted with N_{n+1}/N_n on the ordinate and log N_n on the abscissa.

Broadhead & Wapshere (1966) studied two congeneric species of psocids which live together and feed on algae and fungus on the trunks and twigs of larch trees. The two species differed slightly in distribution—*Mesopsocus immunis* had a small preference for the live branches. The eggs of *Mesopsocus unipunctatus* were attacked especially by a tiny Mymarid wasp of the genus *Alaptus*. The nymphs of *M. immunis* were preferentially attacked by the Braconid wasp *Leiophron*. We shall see in Chapter 4 how specific parasites like these can stabilize the population of each host species independently. Regression analysis of their four years of census results showed no significant relationship between survival and food or the numbers of the other *Mesopsocus* species.

Pontin (1969) was able to demonstrate that ant colonies had direct adverse effects upon each other. He studied two abundant species of field ant, *Lasius niger* and *Lasius flavus* which builds conspicuous mounds. Both species form fairly distinct colonies centred on at least one fertile queens. The difficulty was that colony size could be fully assessed only by a destructive census method. Instead, Pontin counted only the young winged queens produced by each colony. These aggregated under flat stones he provided for each colony. When the stones were lifted the young queens could be removed and counted, and the stones replaced with a minimum disturbance to the colony. Evidence for competition came first from the distribution of colonies in space: they were not randomly distributed. Those which produced many queens occupied a large area. This space was apparently the feeding ground in which the worker ants exploited the root feeding aphids. Pontin showed by experiment that the reproductive success of a colony of *Lasius flavus* or of *Lasius niger* could be increased by removing adjacent colonies, or reduced by transplanting and establishing a thriving nest mound of *Lasius flavus* in close proximity. His results showed that the adverse effect of competition was greater on nearby *flavus* nests than on those of *niger*. These are just the conditions defined in Fig. 3.4D for stable co-existence, which is generally observed in these two widespread species of ant.

3.7 Competition for carrion

Competition is exceptionally severe for nutritious food material such as carrion. It is quickly exploited by specialist scavengers amongst many vertebrate and invertebrate groups and the outcome depends critically on timing. Generally the advantage lies with the first arrivals and blow-flies such as *Lucilia* and *Calliphora* often assemble around a dying mammal and begin to lay eggs at about the time of death. The sheep blow-flies, *Lucilia* spp., of Australia and Europe have gone further and lay eggs on healthy sheep. The maggots produce enzymes which digest the skin of the sheep and the resulting lesions grow rapidly and soon

kill the sheep. The maggots, together with some later arrivals, compete
for the corpse. Fuller (1934) described the complex ecological succession
which develops.

Springett (1968) studied the competition between blow-flies and
burying beetles (*Necrophorus*) for the corpses of mice under carefully
controlled conditions. The outcome was influenced both by the timing

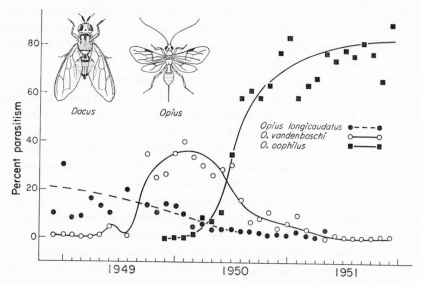

Fig. 3.11 Parasites of the fruit-fly *Dacus dorsalis* successively established in Hawaii for biological control of this pest provide an illustration
of competitive elimination. Data from Bess *et al.* (1961).

of the introductions and by the presence or absence of mites of the
genus *Poecilochirus*, which are usually carried about on the bodies of
the adult beetles. In the absence of the mite, a pair of beetles was
unable to produce any offspring when in competition with either 100
eggs or 100 larvae of *Calliphora*. If the beetles carried 30 mites between
them, the mites destroyed all the blow-fly eggs, and both beetles and
mites thrived. The mites were unable to destroy *Calliphora* larvae,
which then prevented either the beetles or the mites from breeding.
It is interesting to have proof that the commonly observed association
between the mites and the burying beetles is mutually advantageous.
We might conclude from Springett's observations that the success
in competition for carrion is largely a matter of timing; but if the

disadvantage of late arrival is simply related to delay in exploitation, then we would expect in any region the species most usually first would replace the other. But although competitive displacement is observed in the single corpse, both species are widespread and abundant in Europe. Other factors must also be important, perhaps related to season, or to the size of a corpse and its situation.

3.8 Field observations of competitive displacement

Competitive displacement of one species by another can best be ob-served as the result of field experiments. A good example comes from the introduction into Hawaii of a number of braconid parasites of the genus *Opius* (Bess *et al.* 1961) (Fig. 3.11). The objective was the biological control of the fruit-fly *Dacus dorsalis*. First *Opius longi-caudatus* was introduced and established but the percentage of *Dacus* parasitized remained low. When *Opius vandenboschi* became estab-lished it virtually eliminated *O. longicaudatus*, but then *Opius oophilus* rapidly increased in numbers and caused a much higher percentage of parasitism than previously recorded and the two other species were both eliminated.

DeBach & Sundby (1963) describe in some detail the fate of succes-sive introductions of parasites of the genus *Aphytis* which are parasites of the red scale, *Aonidiella aurantii* on citrus (Fig. 9.3c). Their story involves competition between congeneric parasite species and tells how different species reacted to the very different climatic conditions in various citrus growing areas of Southern California.

3.9 Discussion

The Lotka–Volterra equations provided a useful line of thought about the way in which species might interact, and the verbal definition of the conditions for co-existence or elimination hold good. Competitive exclusion occurs when the successful species has a more adverse effect upon the other species than upon itself. For co-existence each species must act more severely on itself than on the other species with which it competes. We shall see in Chapter 4 that the co-existence of parasite species which compete for the same host species is helped by 'mutual interference', which is an intraspecific density dependent effect. Mathe-matically inclined readers may amuse themselves trying to incorporate non-linearity into a new theory. This will be more realistic than the Lotka–Volterra equations. However, we saw in the previous chapter, the outcome of a model including non-linear components could be changed in various ways. The outcome of models including two competing species will likewise be complex—too complex to discuss here.

However, these non-linear density dependent effects and the similar interspecific effects which we have found are likely to have importance in genetics as well as in population theory. The elimination of an unfavourable gene is comparable to the competitive exclusion of a rival species in population dynamics. The widespread presence of genetic polymorphism is analogous to the co-existence of competing species. Furthermore the same mortality factors which cause the population changes which we have discussed in this chapter must be acting as selective agents in genetics. Their precise relations with population density or gene frequency may be of critical importance to our understanding of both population and evolutionary problems.

CHAPTER 4

PARASITES AND PREDATORS

4.1 Synopsis

The simple mathematical theories of Thompson and of Nicholson for parasite–host or predator–prey interactions consider the special case when a specific enemy and its prey have discrete synchronized generations. They thus consider only a fraction of the naturally occurring interactions.

The main difference between the theoretical ideas is that they have placed particular emphasis on different features of the interaction; on the number of eggs which the enemy can produce, or on its searching efficiency and the way this is affected by changes in the density of the prey or of the enemy. We suggest that elements from all the theoretical ideas are' necessary to any general model for this kind of interaction. It is also important to include other features in a model. These theories also make the mathematically convenient assumption that the enemy searches for its prey at random, whereas we know from both laboratory experiments and field observations that this is not necessarily true. Parasites and predators tend to aggregate where prey densities are higher. Also, the theoretical models examine the situation where the only factors affecting both the populations are the densities of one another's populations. Natural interactions involve other factors as well—which may have a greater effect on the stability of the system than any of the components included to describe parasite behaviour.

4.2 Introduction

In the experiments we have discussed so far, cannibalism and predation often played some part in inter- and intraspecific competition. But in

c

these cases animal food was not the primary source of energy for the species. In this chapter we shall consider the types of interactions which occur when one species of insect feeds on another species, and the theories which seek to describe them.

Parasitic insects (sometimes termed parasitoids) are really a special form of predator that usually require one host for complete development. They differ from true parasites in the strict zoological sense since they almost invariably kill their hosts. Nevertheless, they are commonly referred to as 'insect parasites' and we shall use this term for them. Most species of insect parasites belong to either the Diptera or Hymenoptera. The adults lay their eggs on, in or near some stage of their host on which the developing larvae feed. Some parasites—and perhaps the majority—show a preference for one or just a few host species, but others are more polyphagous and attack a much wider variety of host species. Some parasites may have more than one generation in a year and each generation may attack quite different species of host.

Predatory insects differ from these parasites in requiring more than one prey individual to complete development and thus the searching for prey is a process which is continued by a series of different growth stages of increasing size.

We know from practical experience of 'biological control' (Chapter 9) that the introduction of parasite or predator species into an abundant insect pest population can result in a rapid build up in the numbers of these natural enemies. This then causes a rapid decrease in the pest population and a subsequent decline in the natural enemy numbers leading to co-existence of both at much reduced densities. Various mathematical models have been proposed to explain how natural enemies may act on prey populations. We shall trace the development of these ideas, and show how the failure of earlier models to provide satisfactory explanations has led to the development of more complicated and more realistic models. All the workers whose ideas we shall discuss here have provided models only for the simplest possible type of interaction, where a specific and synchronized natural enemy attacks a species which has discrete generations and where each individual host or prey found adds a corresponding number to the next parasite or predator generation.

These assumptions are probably more valid for insect parasites than for predators and most of the theoretical models discussed in this chapter were developed for, and are appropriate to, host–parasite interactions. However, even simple host–parasite interactions are not easy to model. The parasite's reproduction depends on the female's ability to find hosts, and thus the mathematical model must include a description of the searching ability of the female parasites. This, of course, means attempting to express features of the behaviour of adult parasites by mathematical formulae.

4.3   W.R.Thompson's models

Thompson (1924) was interested in the possibility of introducing parasites into pest populations. He was thus thinking about what might happen if a relatively small number of parasites were introduced into a vast population of a pest. He thought, perhaps rightly, that under such circumstances the parasites would have no difficulty in finding hosts, and that the parasite's rate of increase would be limited only by the female parasites' egg supply.

He first proposed that the parasites would lay only one egg in each host found. On this basis the number of eggs laid by the parasite population (P_E) would be the average egg complement of a female parasite (a constant C) times the number of female parasites searching (P); that is:

$$P_E = CP \qquad (4.1)$$

Thus, Thompson equated the number of eggs laid with the number of hosts parasitized, but he realized that this assumption certainly would not fit all parasite–host situations. Many parasites are unable to distinguish between healthy hosts and those already parasitized, and in these cases the host may contain more than one egg even when the host can support the development of one parasite only.

Thompson got over this difficulty by assuming that the encounters (N_a) between parasites and hosts were distributed at random, and he used a random distribution formula to calculate the number of hosts attacked (N_{ha}) in these encounters. The number of attacks N_a is equal to the number of parasite eggs laid (P_E) if a single egg is laid at each encounter, so his model for parasitism can be written

$$N_{ha} = N \left[1 - \exp \left(-\frac{N_a}{N} \right) \right] \qquad (4.2)$$

where N is the host density and (P_E) is obtained from equation (4.1).

This model has not provided a good explanation of parasite–host interactions. Certainly, the calculated parasite populations initially increase very rapidly and this has been seen to happen in the field; but, depending on the relative rates of increase chosen for the parasite and host, either the host and parasite populations continue to increase indefinitely or the parasite reduces the host population to extinction and then itself dies out. The indefinite increase of both populations is

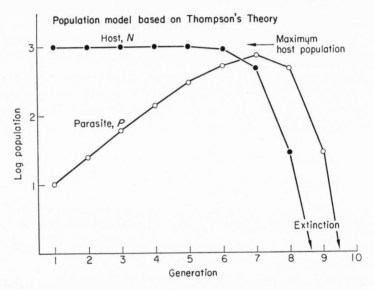

Fig. 4.1 Population model for a stabilized host population attacked by a parasite using Thompson's theory. Host reproductive rate $F = 2$, population stabilized by a density dependent factor at 1000. Parasite lays an average of 2·5 eggs. Formulae (4.1) and (4.2) apply.

clearly a practical impossibility. At some density the host population will be limited by the availability of food. When an upper limit is set to host numbers by introducing a density dependent factor into the model, in addition to the parasite, then the usual outcome becomes the extinction of both the host and the parasite as shown in Fig. 4.1. Only if the parasite also suffers a density dependent mortality can the model be stable. The host is then regulated by the density dependent factor and the continued existence of a small parasite population fails to affect host numbers.

Although Thompson's ideas might correctly express the initial relationships when a few parasites enter a region of abundant hosts, they are clearly not a good description when the hosts are comparatively rare and parasites' *searching* ability must become important; nor do they provide an explanation for the co-existence of the parasite and host populations at new reduced levels.

4.4 A.J.Nicholson's models

Nicholson (1933) and Nicholson & Bailey (1935) considered quite a different situation from that which interested Thompson; one called the 'steady state' where the parasite and host populations co-exist in a state of equilibrium. Essentially Nicholson had two models. The first was a purely verbal one in which he agreed with Howard & Fiske (1911) and with Smith (1935) (see Chapter 2) that populations were regulated by factors which acted in a density dependent way. However, in his mathematical model for a parasite–host interaction, the parasite did not act in a density dependent way.

Nicholson nevertheless thought of parasites as being regulating factors. He assumed that parasites would search for their hosts at random, and that their rate of increase would be restricted not by their egg supply but by their ability to find hosts. Thus, he assumed:

1 That the rate at which parasites find hosts is proportional to the host density. This presupposes that the parasites are never limited by their egg supply.

2 That the average area which one parasite searches in its lifetime is constant and characteristic for that species. This he called the parasite's *area of discovery* which is represented by the symbol (a).

On the basis of these assumptions Nicholson produced his 'competition curve' (Fig. 4.2D) in which the percentage parasitism rises asymptotically towards 100 per cent as the parasite density increases. The mathematical description of this curve is the basis of all the Nicholson–Bailey models, and because these ideas about the way in which parasites search have been incorporated into the models of other workers, we must explain them in a little more detail.

Figures 4.2A and 4.2B represent schematically the search by an individual parasite (A) and by a parasite population (B) as assumed by Nicholson. The track of a parasite searching on a flat surface is shown in Fig. 4.2A; the width of the track is determined by the parasite's ability to detect hosts on either side of its line of movement. The length of the track represents the distance the parasite moves whilst searching throughout its lifetime, and the total area searched is the

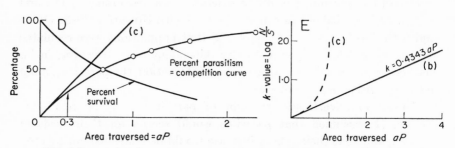

Fig. 4.2

A Hypothetical track of randomly searching parasite.

B Schematic representation of 30 parasites searching in a given area. Each square represents the area of discovery (0·01 of total area) of a single parasite.

C As B, but parasites search systematically by avoiding areas previously traversed.

D Nicholson's 'competition curve' (b). The straight line (c) would be obtained if the parasites searched systematically.

E As D, but parasitism expressed in k-values to illustrate derivation of equation (4.3)

area of discovery (a) of the parasite. The movements of the parasite are random in that they are unrelated to host distribution and the parasite can recross its tracks and thus search areas which have already been searched. Because the parasite can recross its tracks, the area effectively searched is less than the area traversed.

Figure 4.2B shows a diagram of the searching by a parasite population of 30 individuals. For convenience, the area of discovery of each parasite has been drawn as a square of 0·01 of the total area. The positions of the squares have been determined at random which means that no individual parasite avoids those areas which have already been searched by other parasites. It will be clear from this that the amount of overlap of the areas searched must increase as the number of parasites is increased. In this example the total area transversed by the parasites (aP) is 0·3 of the total area, but the area which was effectively covered is only 0·254 of the total area. In other words, we may expect only 25·4 per cent of the hosts to be parasitized instead of the 30 per cent which would have occurred if there had been no overlap of searching by the parasites (Fig. 4.2c). An examination of Fig. 4.2D shows that this value of 25·4 per cent is very close to that expected from the 'competition curve' which is 25·9 per cent (see arrow). The difference arises from random error and is not significant.

These examples should have made it clear that there is nothing uniquely biological about Nicholson's mathematical formulae. His basic assumptions are:

1   no egg limitation;
2   search is random; and
3   the area of discovery a is constant.

Provided these are correct, then his predictions about the consequences of a parasite–host interaction must logically follow.

If we are to test this model we must estimate the area of discovery from field or laboratory data. The area of discovery is easily calculated if we can measure the number (or density) of parasites searching (P) and the proportion of their hosts that they parasitize.

$$a = \frac{1}{P} \log_e \frac{N}{S} \qquad (4.3)$$

where N is the number (or density) of hosts exposed to attack and S the number (or density) not parasitized. Figure 4.2E shows the competition curve when replotted using k-values to describe the proportion of hosts parasitized. This results in a linear relationship with a slope of 0·4343 ($= 1/\log_e$):

$$k\text{-value for parasitism} = \log \frac{N}{S} = \frac{aP}{2·3} \qquad (4.4)$$

from which equation (4.3) is obtained by rearrangement.

Equations (4.3) and (4.4) describe Nicholson's 'competition curve' and enable us to predict the proportion of hosts parasitized from the number of searching parasites, provided that the area of discovery is known. It is now but a simple step to produce a population model based on equation (4.4).

$$\log N_{n+1} = \log N_n - \frac{aP_n}{2\cdot 3} + \log F \qquad (4.5)$$

$$P_{n+1} = N_{ha} = N_n - \text{antilog}\left(\log N_n - \frac{aP_n}{2\cdot 3}\right) \qquad (4.6)$$

where F is the host reproductive rate, N_{n+1} and N_n represent successive host populations and P_{n+1} and P_n represent successive parasite populations.

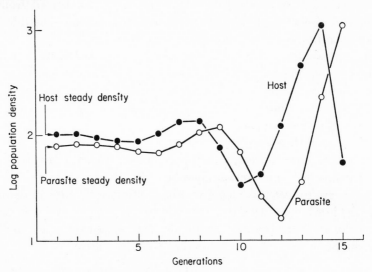

Fig. 4.3 A population model based on Nicholson's theory (equations 4.3 and 4.4).

$\text{Log } N_{n+1} = \log N_n - \dfrac{0\cdot 02P}{2\cdot 3026} + \log 5.$

Figure 4.3 shows the outcome of a parasite–host model based on equation (4.5). There are, according to the model, certain 'steady densities' N_s and P_s of host and parasite at which successive populations are exactly the same size. From formula (4.5), if we put $N_{n+1} = N_n = N_s$ then it follows that

$$\frac{aP_s}{2\cdot 3} = \log F \quad \text{so that} \quad P_s = \frac{2\cdot 3 \log F}{a} \qquad (4.7)$$

If there is no other host mortality, $N_s F - P_s = N_s = P_s/(F-1)$. Substituting for P_s we then have

$$N_s = \frac{2\cdot 3 \log F}{a(F-1)} \qquad (4.8)$$

Thus for any given rates of host increase and of the area of discovery of the parasite we can calculate the steady densities of both host and parasite. If either host or parasite are displaced from this steady density the model produces increasing oscillations in the populations of both species. The steady density represents an unstable equilibrium in this simple model. We shall see later how it may be stabilized.

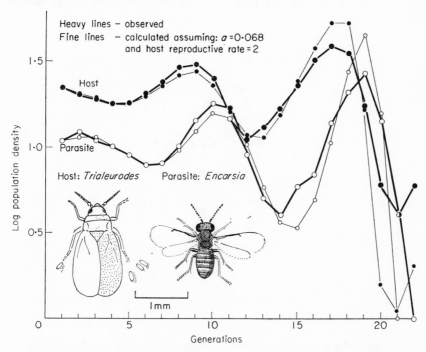

Fig. 4.4 Observed and calculated results of an interaction between *Encarsia* and the greenhouse white-fly, *Trialeurodes*. Model calculated on basis of a constant area of discovery of 0·068 and a host reproductive rate of 2. After Burnett (1956).

Several workers have carried out laboratory experiments to test if parasite–host interactions in isolated conditions are of this form. The heavy line in Fig. 4.4 shows the results of an experiment carried out by Burnett (1958) using the greenhouse white-fly *Trialeurodes vaporariorum* and its chalcid parasite *Encarsia formosa* in an interaction which simulated 22 generations. These generations were artificial since 'the average number of hosts parasitized, rounded to the nearest whole number, in one generation gave the initial parasite density for the next.

The average number unparasitized in one generation was doubled and rounded to the nearest whole number to give the number of hosts exposed in the following generation on the assumption that the reproductive rate of the host was two'. Burnett found fairly good agreement between his experimental results and a Nicholsonian model using the average of discovery and a host reproductive rate of two. However, it is noticeable from Fig. 4.4 that the observed interaction was rather

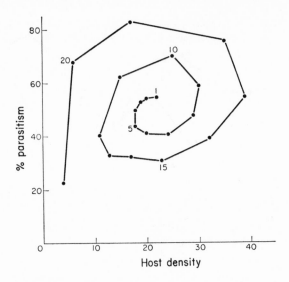

Fig. 4.5 An example of a delayed density dependent relationship using the observed results from Fig. 4.4.

more stable than that predicted by the model (fine line). DeBach & Smith (1941) used the chalcid parasite *Nasonia vitripennis* and puparia of the house-fly as host in a similar experiment, which produced similar results over 7 generations. (See exercise 4.2.)

The oscillations in parasite and host populations generated by Nicholsonian models and observed in these experiments arise because the reproductive rate of the parasites is related to host density. As can be seen from Fig. 4.3, the population density of the parasite increases whilst host density is above the mean, and it continues to increase after that of the host has started to decrease. Conversely, the parasite continues to decrease even when the host has started to increase. If the percentages of parasitism generated by such a model are plotted against the respective densities of the host, the graph obtained is quite different from that of a density dependent relationship. When the points are linked in a time series, an anti-clockwise spiral is produced; this is shown in Fig. 4.5 using data from Burnett's experiment with *Encarsia* and *Trialeurodes*. Clearly, instead of there being a positive relationship between percentage parasitism and host density, as would be expected if parasites acted as density dependent factors (see Chapter 2), the relationship varies between being directly and inversely density dependent with periods in between when it is approximately independent of density. The relationship is density dependent when both populations are rising or falling together, and inverse when the host is increasing whilst the parasite is decreasing, or vice versa. We can thus state quite firmly that the effect of a Nicholsonian parasite is very different from that of a density dependent factor: the tendency of a density dependent factor is usually to stabilize, the tendency of a Nicholsonian parasite is to produce population oscillations of increasing amplitude—a kind of instability. Because of this delayed relationship between percentage parasitism and host density Varley (1947) coined the term *delayed density dependent factor* to describe the effect of a Nicholsonian parasite.

Nicholson himself was fully aware that increasing oscillations do not occur under natural conditions, and he suggested that the natural outcome of increasing oscillations would be to cause the population of the host to fragment and continue to exist only as small and widely separated subpopulations. He envisaged the increasing oscillations occurring in the sub-populations, but because these would not be in phase with one another, some populations would become extinct whilst others would be started up in new places by immigrants. This sort of pattern perhaps occurs in the interaction between the moth *Cactoblastis cactorum* and its food plant, the prickly pear, in Australia (see Section 9.3), but such events have not been sufficiently documented in field studies.

There is no need to reject Nicholson's ideas just because in their simplest form his models generate increasing oscillations in numbers. We saw in Chapter 2 that stability can be brought about by the action of an appropriate density dependent factor. Any Nicholsonian model may be stabilized by introducing a density dependent mortality to act either on the host or on the parasite population (or on both) providing it is of sufficient strength.

There are, however, other difficulties with Nicholson's ideas about

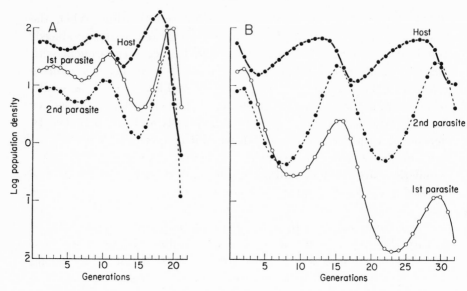

Fig. 4.6

A   A Nicholsonian model with one host and two parasite species acting successively (after Nicholson & Bailey (1935), Fig. 12).

B   As above but an additional density dependent mortality (k-value = $b \log N_n$) now acts on the hosts surviving parasitism.

$$\log N_{n+1} = \log N_n - \frac{a'P'}{2 \cdot 3026} - \frac{a''P''}{2 \cdot 3026} - b \log N_n + \log F$$

(where $a' = 0 \cdot 025$; $a'' = 0 \cdot 035$; $b = 0 \cdot 15$; $F = 2$).

parasite–host interactions. For example, under natural conditions it is quite common to find an insect species being attacked by two or more specific and synchronized insect parasites, which may attack either the same or successive stages of their host. Nicholson clearly thought that his models allowed for such parasite co-existence, and in Fig. 4.6A we show a recalculation of the graph used to demonstrate this supposed co-existence (Nicholson & Bailey 1935, Fig. 12, p. 588). However, the co-existence is illusory: the tendency of one species to replace the other is hidden by the rapidly increasing magnitude of the oscillations. Figure 4.6B shows a similar calculation, but a little extra stability has been put into the model in the form of a weak density dependent factor acting on the host. In this model the first parasite is clearly being eliminated by the second. If parasites are assumed to act

in the way postulated by Nicholson, it is virtually impossible to account
for parasite co-existence unless we suppose that each parasite species also
has a density dependent factor which acts only on its own population.

Probably the most serious objection to Nicholson's ideas is to the
assumption that the average area of discovery of a parasite is constant
for a given species. The areas of discovery of *Encarsia* and *Nasonia* in
the experiments described above were fairly constant, but field studies
have shown how important weather conditions can be in determining
searching efficiency and there is now a large amount of data from
laboratory experiments showing the search efficiency to be dependent
on both host and parasite density.

4.5 The effect of host density

Solomon (1949) defined the term 'functional response' to describe
changes in the **number** of attacks per parasite (or predator) as host
(or prey) density changes. This response has been extensively studied
by Holling.

Holling (1959) demonstrated that a completely general feature of
parasite or predator attack—the 'handling time'—has a very important
effect on the functional response. This handling time is the interval
between a natural enemy first encountering a host or prey and search
being resumed. It varies considerably from species to species. For
example, Hassell & Rogers (1972) found that the ichneumon *Nemeritis
canescens* spends an average of about 20 seconds between first encountering a host and resuming its search. On the other hand, the handling time
of *Nasonia vitripennis* depends on the previous history of the female
parasite (Varley & Edwards 1957) and may be several hours if the
female is hungry or immature. The proportion of the total time spent
in handling hosts must increase as more hosts are found (i.e. as host
density increases), and this reduction in the time spent searching at
high host densities must reduce searching efficiency (Holling 1959,1966).

A typical functional response is shown in Fig. 4.7; expressed in
terms of changing **numbers** attacked in Fig. 4.7A and **proportion**
attacked in Fig. 4.7B. It is clear from Fig. 4.7B that from the point
of view of the hosts this response to host density is inversely density
dependent. (The dotted lines show the expected response if searching
efficiency were independent of host density as assumed by Nicholson.)
The handling time alone may limit the maximum attack rate per

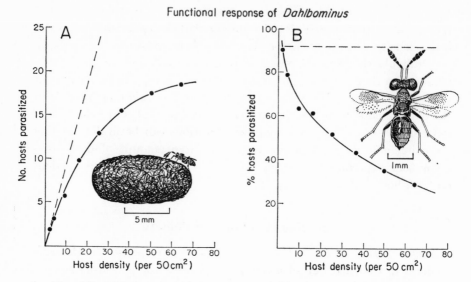

Functional response of *Dahlbominus*

Fig. 4.7 The functional response of a single female of the chalcid parasite *Dahlbominus fuscipennis* searching for cocoons of the saw-fly *Neodiprion sertifer* within cages of 50 cm floor area.

A Expressed as the number of hosts parasitized at different host densities, and

B Expressed as the percentage of hosts parasitized at different host densities.

(The broken line shows the response expected of a Nicholsonian parasite). Data from Burnett (1956).

parasite at high host densities. In many instances, however and particularly when the handling time is short, egg limitation of parasites, or satiation in the case of predators may be more important. The effect of handling time on the number of hosts encountered (N_a) at different host densities was studied by Holling (1959) by an experiment in which the searching of an insect natural enemy was represented by the finger of a blindfolded person searching on a table for sand paper discs. He fitted equation (4.9) to the results and it is commonly referred to as the 'disc equation'.

$$N_a = \left[\frac{Ta'N}{1 + a'T_hN} \right] P \qquad (4.9)$$

where T is the total time available for search, N the initial number of prey available, T_h the handling time and a' the coefficient of attack.

This equation only predicts the number of hosts parasitized (N_{ha}) if search is systematic (Royama 1971, Rogers 1972). If search is random, N_{ha} may be predicted by substituting equation (4.9) in (4.2). This now provides a better submodel for parasitism. The consequences of such a model are similar to a Nicholson–Bailey model but the outcome is always somewhat more unstable, because the functional response is inversely density dependent.

4.6   The effect of parasite density

Searching parasites may change their behaviour if other individuals of the same species are nearby or after they have detected a parasitized host. Such behaviour tends to result in a decreasing searching efficiency as parasite density increases. Hassell (1971a, b) investigated some features of the behaviour of *Nemeritis* when attacking its host *Ephestia cautella*, a species of flour moth. He found that when two searching parasites met, one or both of them tended to leave the area where the encounter took place. This interference between parasites, which detracts from their searching efficiency, must increase as parasite density increases.

Hassell & Varley (1969) examined the published data from a number of laboratory host–parasite interactions and found that the relationships between area of discovery and parasite density could all be represented fairly well by the formula:

$$\log a = \log Q - m \log P \tag{4.10}$$

or

$$a = QP^{-m} \tag{4.11}$$

so that if we substitute for a in (4.4),

$$k = \frac{QP^{1-m}}{2 \cdot 3} \tag{4.12}$$

where Q is the 'quest constant' (the area of discovery when the parasite density P is 1 and m is the 'mutual interference constant' (the slope of the relationships in equation (4.10)). Figure 4.8 shows some of these relationships. Nicholson's assumption of a constant searching efficiency is now only a special case of the more general model when $m = 0$. Equation (4.11) is a very convenient description of how searching efficiency changes with parasite density; the parameters can easily be determined from field data provided that the percentage parasitism and the density of searching parasites are known. Notice, however, that equation (4.10) is only an approximate description; several of the relationships in Fig. 4.8 show signs of being curvilinear. It is also

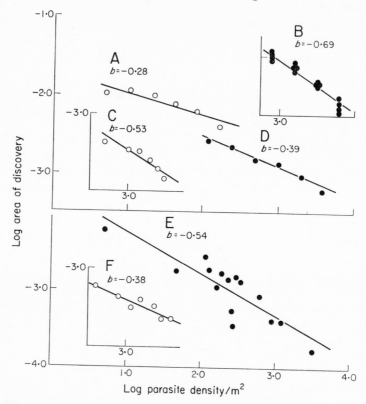

Fig. 4.8 Relationships between log area of discovery and log density of searching parasites.

A *Dahlbominus fuscipennis* (see Fig. 4.7) Burnett (1956).
B *Pseudeucoila bochei*. Bakker *et al.* (1967).
C *Chelonus texanus*. Ullyett (1949a).
D *Encarsia formosa* (see Fig. 4.4). Burnett (1958).
E *Nemeritis canescens*. Hassell & Huffaker (1969).
F *Cryptus inornatus*. Ullyett (1949b).
For references see Hassell (1971a).

obvious that searching efficiency cannot indefinitely increase as parasite density is reduced.

The outcome of some models based on Quest Theory can be completely different from that of a Nicholsonian model because they include 'interference'. In the first place, the changing searching efficiency is equivalent to a density dependent factor acting on the parasite so the models are not necessarily unstable. The stability of the models increases with greater values of m as shown in Fig. 4.9. Secondly, there are wide ranges of values for Q and m in equation (4.11) which

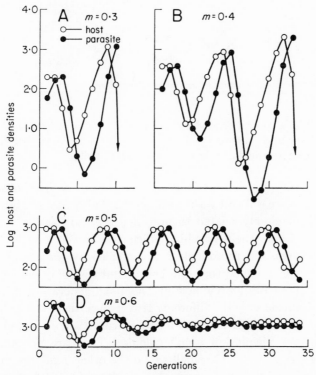

Fig. 4.9 Population models showing the increasing stability as the mutual interference constant m is increased from 0·3 in A to 0·6 in D. From Hassell & Varley (1969).

allow the co-existence of two or more parasite species on a single species of host (Fig. 4.10). These questions of stability and co-existence are very important in developing a theoretical basis for biological control (see Chapter 9).

4.7 Predator–prey interactions

Much of this chapter has been concerned with population models for simple host–parasite interactions. We also need population models for

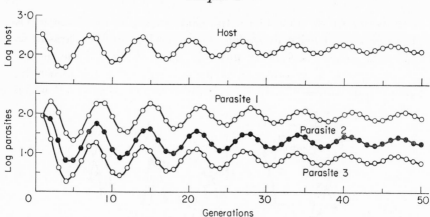

Fig. 4.10 A population model showing the stable co-existence of a host and three parasites. Host $F = 5$; $Q = 0.1$; $m = 0.5$ for parasites 1 and 3; $m = 0.6$ for parasite 2. From Hassell & Varley (1969).

insect predator–prey interactions but deriving these from observation is proving a more difficult task. Perhaps this is because predator reproduction is not closely related to prey density and because it is much harder to estimate their searching efficiency. In the models discussed earlier, the parasite's reproduction is modelled by the attack formula used, since it is assumed that one host found results in one or a given number of parasite individuals in the next generation. The insect predator situation is quite different. Hover flies, such as *Syrphus* spp., need to feed on pollen, nectar, etc. to mature their eggs and tend to oviposit in the vicinity of aphid colonies. They are predatory only during the larval stages. Coccinellids, on the other hand, are predatory as larvae and adults, but again the number of eggs laid is usually independent of prey density provided that sufficient food has been obtained for complete egg maturation. The searching efficiency depends not only on prey and predator density in much the same way as for insect parasites, but also on the particular stage of development of the predator. Dixon (1958) showed this from the relationships between a predatory coccinellid, *Adalia decempunctata* and the nettle aphid, *Microlophium evansi*. The adult and the four larval instars of *Adalia* all feed on aphids. Table 4.1 summarizes some of Dixon's data: the success of predation resulting from encounters between *Adalia* and the nettle aphid depends on the different instars concerned. For a given size of aphid, each successive larval instar of the predator has a better chance

Table 4.1 Percentage chance of successful predation of the nettle aphid *Microlophium evansi* when encountered by the predatory Coccinellid *Adalia decempunctata*—the ten spot lady bird (Dixon 1958).

Lady bird instar	Aphid instar				
	1	2	3	4	Adult
1	19	1	0	0	0·5
2	45	16	2	0	0
3	83	50	45	10	7
4	90	60	43	32	19
Adult	38	26	17	6	11

of success in capturing its prey and thus getting food. The adults are rather less efficient than the big larvae.

We can see from this that some of the simple assumptions which seemed to fit parasite behaviour certainly do not apply to this kind of predator. In the first place, searching efficiency depends on the stage of development (as shown by Dixon) and secondly there is not such a clear relation between reproduction and prey density.

Even with an ideal submodel for predation, we would still require additional information to model the outcome of a particular predator–prey interaction. Predator populations, just like their prey, suffer several mortalities, whether from climatic effects, secondary predation or competition. (This, of course, is also true of host–parasite interactions.) These life table components must be included in any model designed to mimic field conditions. Some examples where this has been attempted are described in Chapters 7 and 9.

4.8 Non-random search

The models we have discussed in this chapter assume that the population of parasites or predators searches at random with respect to their hosts or prey. This assumption of random search simplified the mathematics—but there is little or no evidence that it is generally true, rather the reverse. Many insect natural enemies do not search at random, but show a marked response to the spatial distribution of their hosts or prey. Some predators and parasites are attracted from a long distance by the scent of their host or prey. Others tend to remain for a longer time searching in areas where they have already found a host or prey. In either case there will be a tendency for the searching population to spend more time and therefore to aggregate in areas where their hosts or prey are most abundant and to cause a higher percentage mortality there compared with low density areas (Hassell 1966, Murdie & Hassell 1973). Figure 4.11 shows an example of such a behaviour response for *Cyzenis albicans*, a tachinid parasite of the winter moth (see Chapter 7). The density dependent response to the host distribution is the result

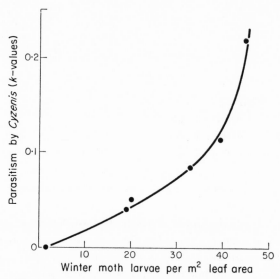

Fig. 4.11 The behavioural response of the tachinid parasite *Cyzenis albicans* to local differences in the population density of winter moth larvae on different trees in one area in the same year.

of the parasites tending to concentrate their searching activities on the oak trees with the highest winter moth densities. Non-random search of this kind is important since it tends to increase the stability of the host–parasite or prey–predator interaction (Hassell & May 1973).

Field studies give the general impression that populations of polyphagous parasites and predators tend to be especially stable. We know that insectivorous vertebrates, especially birds, concentrate on the more abundant host or prey species, and probably insects can do so likewise. Perhaps stability is also promoted by this type of deviation from random search. This ability to change to different food supplies and concentrate on the most abundant could be an important factor in stabilizing predator or parasite populations. This problem is further discussed in Chapter 7.

CHAPTER 5

CLIMATE AND WEATHER

5.1 Synopsis

Weather and climate affect the physiology and behaviour of insects. We include in weather the changing hour-to-hour measurements of temperature, humidity, wind and rain etc. Climate is defined in terms of long-term averages for these same measurements.

Day-length and temperature have important effects on the insect's endocrine system which can act as a switch and determine whether the insect is active or goes into diapause. This modifies the effects of weather factors on survival, development and reproduction.

Weather affects insects as individuals (independently of the population density), and so acts as a *density independent* or *catastrophic factor*. Weather is known to determine population change in many insects.

The vexed question 'can weather control populations' is examined: if a population is acted upon both by a density dependent factor and a density independent variable factor such as weather, then the *weather determines the changes*, but the density dependent factor is primarily responsible for *regulating* the population about its average level of abundance. Confusion arose because the word *control* has been used indiscriminately for both these processes.

5.2 Introduction

The planet Earth spins daily about an axis of rotation which differs by $22\frac{1}{2}°$ from that of its orbit around the sun. This imposes a diurnal rhythm in incident solar radiation and also an annual rhythm. The diurnal rhythm is dominant at the equator and the annual rhythm dominates the polar regions. In temperate latitudes both are important.

The changes in measurable physical properties of the atmosphere, and of freshwater and marine environments affect animals and plants in various complex ways.

5.3 Climate

To define *climate* meteorologists have done two things.

1 They have standardized their temperature measurements by putting their instruments inside a Stevenson screen—a white wooden box with louvred sides at a height of 4 ft above a mown lawn.

2 They have taken average values for the sunshine, rainfall, wind, temperature and other weather conditions recorded. The average temperature of a day is the arithmetic mean of the maximum and minimum values recorded. The average for the month is the average of these figures and so on.

This information can be summarized as isothermal lines on a series of maps. Climatic zones are determined by temperature and the distribution of rainfall, both being greatly influenced by latitude and by altitude above sea level.

The geographical range of a particular species of insect may be restricted to a single climatic zone either because its food plant is so restricted, or because in adjacent climatic zones weather conditions are temporarily unsuitable for its life (Birch 1957). Casual immigrants, or their offspring, perish. But in practice distribution cannot be described in terms of climate alone, as can be seen from the very complex and irregular patterns of butterfly distribution in Europe (Higgins & Riley (1970)).

Nevertheless, the species composition of the insect fauna found in interglacial peat deposits is being used with success to indicate the climatic changes during these periods (Coope 1970).

5.4 Weather and its effects

The annual, seasonal and diurnal changes in temperature, humidity, rainfall, snow storms, hail, wind and sunshine constitute the weather. Descriptions of the weather must concentrate on variations and divergences from the mean for the time of year (which is the climate by definition).

Instruments can record many of these environmental variables in a form suited to the computer. The computer can then be programmed to seek correlations between population figures and the meteorological measurements. The problem is how to integrate the rapid environmental changes over the period of days or months needed for a developmental stage or the life of an insect to be completed. The difficulties are increased by the ability of the insect to move and select suitable conditions from a diverse environment. Wellington (1957) has reviewed ways of estimating the nature of this diversity in relation to Stevenson screen measurements for any specific type of synoptic situation.

Without an intimate knowledge of the physiology and behaviour of the insect itself, it is hard to know what to tell the computer to do with the measurements of weather. The weather can produce physiological effects on insect populations in four major ways, by modifying (1) the activity of the endocrine system, (2) survival, (3) development and (4) reproduction.

5.4.1 *Effects on the endocrine system*

The effect of weather on the endocrine system is so basic that we must consider this first, because it acts rather like a switch. Insects of the temperate zone tend to react to decreasing day length by a change in the neurosecretory cells of the brain which induces *diapause* (Lees 1955), a state in which the insect's reaction to temperature is switched off, or at least suspended. The insect may become cold-hardy (i.e. resistant to freezing) and cease to develop or to reproduce. Two known ways of terminating diapause are by increasing day-length, as in the dragon-fly *Anax* (Corbet 1956), or by a period of low temperature, with conditions which simulate winter. Activity is then switched on again by the endocrine system and the insect once more reacts to temperature through the process of growth and development.

Turnock (1973) showed that in Manitoba weather conditions cause a great variation in the diapause and indirectly in the survival of the tachinid fly *Bessa harveyi* which is a parasite of the larch saw-fly, both of which normally have an annual life cycle. If the season is an early one, the newly emerged larvae of *Bessa* are subjected to high temperature and a long day length. They then fail to diapause and emerge quickly as a second adult generation at a time when the acceptable stage of the favoured host is absent. This lack of synchrony causes heavy mortality. Turnock found a strong negative correlation between autumn emergence in consecutive years, which suggests that a genetic factor may also be involved in the timing of the life cycle of *Bessa*.

5.4.2   *Effects on survival*

Direct evidence that any dead insect found was killed by weather factors is hard to obtain although there are many anecdotal records. Frost may kill insects; dead insects can be found on the mountain snow

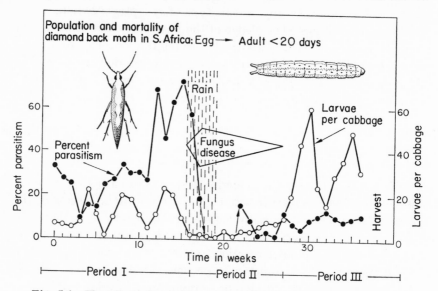

Fig. 5.1   The effect of parasitism and a fungus disease on a population of the diamond back moth, *Plutella maculipennis*. Data from Ullyett (1947).

fields in spring. After heavy rain the water which accumulates in the leaf bases of the teazle (*Dipsacus fullonum*) often contains drowned aphids. Summer floods drowned a proportion of the pupae of the knapweed gall-fly (see Table 6.3) and unusual winds may carry migrating insects like locusts far out to sea, where all perish either by drowning or by being consumed by the marine fauna.

Important indirect effects associated with weather changes are probably frequent, but difficult to find in the literature and we will mention only two.

The diamond back moth, *Plutella maculipennis* is a widespread pest of cabbage and other Brassica crops. Ullyett (1947) provided figures for population and mortality changes in S. Africa (Fig. 5.1). When the rains came, conditions became suitable for the rapid transmission of a fungus disease of the caterpillars and a high proportion of them died. The wet weather also had an adverse effect on the activities of the parasites of *Plutella* so that the percentage of larvae parasitized fell rapidly, which led to a sharp fall in the average parasite population. Another instance, mentioned again in Section 9.4.3 concerns the effect in Fiji of the rainy season on the mite *Pyemotes*. In the dry season the mite was able to destroy most of the larvae and pupae of the coconut beetle *Promecotheca* (Fig. 9.3B), which is a leaf-miner living in tunnels in coconut leaves. In the rains the mites did not survive and the beetles then increased in numbers.

5.4.3 *Effects on development*

Temperature has a simple effect on many chemical reactions. Van t'Hoff and Arrhenius used different formulae to describe these effects, but over the narrow range of temperature in which most insects live

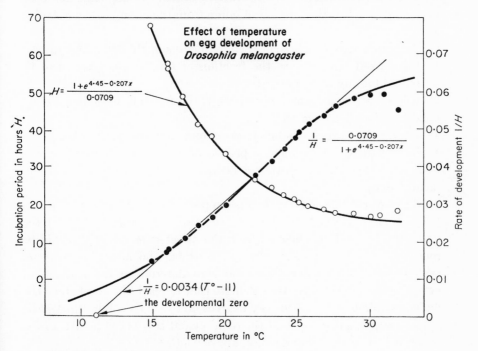

Fig. 5.2 Data from Davidson (1944).

they differ very little and neither of these formulae is very helpful
because at temperatures near freezing and at temperatures above 30°
or 40°C normal physiological processes are hindered and few insects
can live very long.

Davidson (1944) reviewed the effects of constant temperatures on
the development of insect eggs. The duration of the egg stage of the
fruit-fly *Drosophila* (Fig. 5.2) has a minimum at 30°C, so that the curve
for its reciprocal, the rate of development, has a maximum at this
temperature. Davidson fitted a logistic curve to the rate figures, but
this deviates from the measured values at temperatures above 29°C.
It is simpler and accurate enough for our purposes to fit a straight line
to the graph of developmental rate against temperature. This straight
line cuts the temperature scale at the 'developmental zero' (11°C in Fig.
5.2); and at temperatures between 15–27°C, the rate of development
is given, to a first approximation, by the formula

$$1/H = 0 \cdot 0034\,(T - 11) \qquad\qquad (5.1)$$

where T is the constant temperature at which development has been
measured and H is the number of hours from the laying to the hatching
of the egg.

There are four simple logical consequences of this relationship:
1 for a fixed temperature the fraction of total development can be
estimated for any time interval;
2 when mean daily temperatures vary, the fractions of daily development can be added together;
3 when this total equals unity, development is complete;
4 for complete development a precise number of 'day-degrees' needs
to be accumulated (temperature being measured above the developmental zero).

Shelford (1927) used day degrees to predict the time of events in the
life history of the codling moth, *Cydia pomonella*, and found that the
dates at which adults emerged from the pupal stage in spring could be
related to the daily meteorological records. The mean temperature
each day above the developmental zero determined the fraction of
development in each day. We calculate that to reach the adult stage the
insect required about 650 day-degrees above 11°C. For *Drosophila
melanogaster* egg development the requirement is $1/0 \cdot 0034$ hour-degrees,
which is just over 12 day-degrees, above 11°C.

The rate of development of an insect species may change with climatic trends caused by latitude or altitude in a way which seems easy to understand. A widely distributed pest like the codling moth has a single generation each year in the northern part of its range. Further south it may get through two or even three generations in the year. Similarly, in any one place many insect species pass through an extra generation in a year when the weather is exceptionally warm.

The fascinating story told by Lloyd & Dybas (1966) about the American cicadas is much less easy to explain. *Magicicada septemdecim* takes 17 years to grow from egg to adult in the northern part of its range in eastern USA and is famous for the synchronized emergence of the noisy adults. Lloyd found with these cicadas some rather smaller ones whose song is distinctive and which are also sufficiently different in ecology and structural detail to be regarded as distinct species, *M. cassini* and *M. septendecula*. In any one place almost all the individuals of all three species emerge in the same year; nevertheless populations in different places come out in different years, but always at 17-year intervals. The exception is that, in the warmer southern states, the 17-year cicadas are replaced by what appear to be races of the same three species which produce adults at intervals of 13 years. What significance, if any, resides in the fact that the favoured intervals are prime numbers, and why no races appear to have life cycles of 12, 14, 15, 16 or 18 years remains a mystery.

The simple relationship used by Shelford between the developmental rate and temperature of the codling moth worked fairly well for events in the spring, but Hopkins (1918) had already found that it did not hold at all well for the whole year. He enunciated a 'bioclimatic law' which fitted the phenological events in the USA far better:

'Other things being equal, the variation in the time of occurrence of a given periodic event in life activity in temperate North America is at the general average rate of four days to each degree of latitude, five degrees of longitude and 400 ft of altitude later northward, eastward, and upward in the spring and early summer, and the reverse in the later summer and fall'.

Hardwick (1971) gives the peak dates for the capture of various common moths at light traps in Saskatoon (Sask., Canada) and in Dayton (Washington, USA) which is over 600 miles S.W. In Fig. 5.3 we have superimposed his results on a diagram which gives an idea of the differences in temperature at the two places and of the changes in day length. From the bioclimatic law we would expect events in

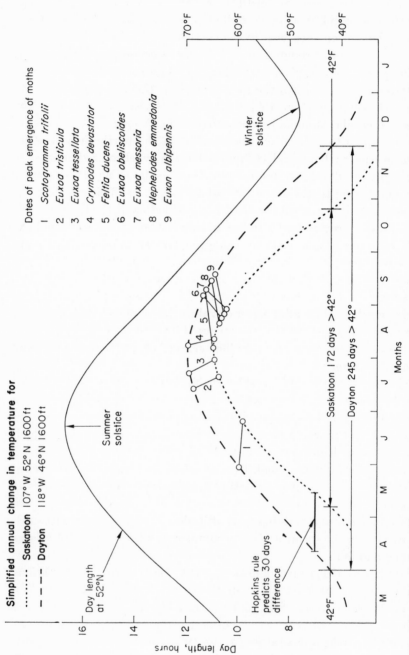

Fig. 53 The relationships between daylength and date of peak capture of nine species of moth as shown by information from Saskatoon (Canada) and Dayton (Washington, USA). Data from Hardwick (1971).

Saskatoon to be a month later than in Dayton in the early spring. The earliest moth listed, *Scotogramma trifolii* peaked on 29th May at Dayton and 26 days later on 24th June in Saskatoon. If temperature alone determines the developmental rate, the delay should be greater for the later species. In fact the peak times for the species in the middle of the summer differed little and the five species which peaked in September in Dayton, peaked in August in Saskatoon, as predicted by Hopkins. The bioclimatic law describes the differences reasonably well, but how are we to explain them physiologically? Perhaps each species has genetically different races in the two places which have evolved different responses to temperature: such a mechanism would be likely to fail for autumn species in cold years—they might be unable to complete their development. More probably, the falling day length after the summer solstice brings about endocrine changes resembling those associated with diapause. If this is the explanation, then development in the new physiological state will have to be accelerated at low temperature and slowed at high temperature.

Sensitivity to day length may be restricted to a particular stage of development. Corbet (1962) showed that the dragon-fly *Anax imperator* normally went into nymphal diapause over the winter. In this state it could be active and feed, but was unable to metamorphose into the adult stage. Just before the penultimate moult the nymph was sensitive to day length. Summer conditions with constant or decreasing day length put the insect into diapause. Development to the adult stage was initiated either by a gradual increase of day length, as normal for March and April in the North temperate regions, or could be induced artificially by a sudden increase of at least 20 minutes day length.

As a consequence of these reactions, if the food supply enabled the nymph to reach the penultimate stage well before the summer solstice, *Anax* developed rapidly from egg to adult in just about a year; if it reached the critical stage after the solstice it entered diapause and was not triggered into development for another nine months. Adult emergence was at almost the same calendar date whether development had taken one year or two.

5.4.4 *Reproduction*

A direct demonstration that temperature and humidity affect the reproductive rate (and also the survival) of an insect is provided by the work of Buxton & Lewis (1934) on the tsetse-fly *Glossina tachinoides* in West Africa. They found that temperature and humidity in the field were often far from the optimum for the species and indeed approached close to the lethal limits. In one series of experiments they determined the thermal death point for one hour's exposure and for one day's exposure. In other experiments at a constant temperature of 30°C and at five different relative humidities between 11 per cent and 88 per cent R.H. the flies were kept in small jars and they were given a chance to feed once a day on human blood. The weight of blood taken at each meal was estimated from the change in weight of the fly. In favourable conditions the female flies, which are viviparous, gave birth to a fully grown larva every few days and any such births were recorded.

Table 5.1 The effect of constant temperature and humidity on the survival, feeding and the birth of the tsetse-fly *Glossina tachinoides*. From Buxton & Lewis (1934).

Temperature in degrees C	Percent relative humidity (%)	Survival in days	Weight of blood meal in mg	Births per hundred ♀ days
30	11	< 10	6	0·3
30	19	20–25	7	0·9
30	44	25–30	11	2·8
30	65	< 5	10	0
30	88	< 5	2	0
24	11	—	—	< 1
24	44	—	—	< 1

Results from this study are summarized in Table 5.1 and Fig. 5.4. Survival of the flies at 30°C was best at 19 per cent and 44 per cent R.H. and was reduced at 65 per cent to less than five days. Both the number of births and the weight of blood taken peaked at 44 per cent R.H.

Superimposed on temperature/humidity co-ordinates of Fig. 5.4 are the average figures (taken from Stevenson screens) for the change in shade temperature and humidity in April (which is typical of the dry season) and July (the season of rains). In July the average conditions for a day are those which, if held constant, would permit a survival of *Glossina* for only five days. The April conditions passed through the

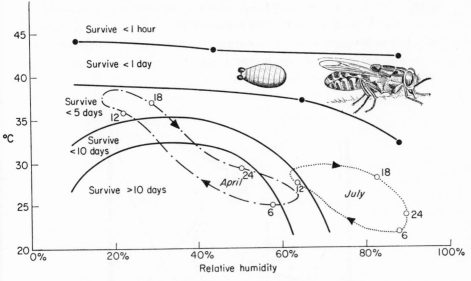

Fig. 5.4 The effects of relative humidity and of temperature on the survival of adult tsetse flies, *Glossina*, in the laboratory, and the average conditions prevailing in West Africa during the months of April and July. Data from Buxton & Lewis (1934).

optimum for the tsetse-flies night and morning (all in Table 5.1), but in the afternoon were again in the five-day survival zone. These measurements show the survival value of changes in tsetse behaviour during the day, and the advantage in the heat of the day of resting under the boughs of large trees where the temperature is well below that of the Stevenson screen.

In the dry season the Stevenson screen temperatures on the hottest days rose to the upper fatal limit for an hour's exposure. The flies' ability to select sheltered places is obviously important to survival. In the wet season conditions for feeding, reproduction and survival are so poor in the field that the tsetse numbers would be expected to fall, as they were observed to do. Indeed Buxton & Lewis's work leaves us with the opposite problem—how the species survives the unfavourable period. This it apparently does in the pupal stage in the ground.

Varley (1947) found a very different relationship between weather conditions and the reproductive activities of the knapweed gall-fly (Chapter 6). At constant temperature and humidity, optimum conditions for egg laying occurred between a temperature of 30–35°C and an

R.H. of 50–70 per cent. Field temperatures never reached this optimal zone. The highest temperatures recorded in the field were associated with low humidities, so the flies were active and laid eggs early on sunny days but hid away as the temperature rose above 20°C and sought cooler and more humid regions near ground level.

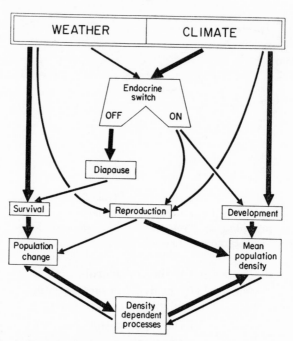

Fig. 5.5 A diagrammatic representation of the main effects of weather and of climate on an insect population.

The four effects of climate and weather which we listed at the beginning of this chapter are effects upon the insect at the physiological level. Figure 5.5 summarizes their main interactions. The point we particularly wish to emphasize in this diagram is this: because climate is defined in terms of *average* conditions, it has a big influence on the mean population density of a species, whereas weather, being the day to day change, includes the *extremes* which have particular influence on survival and hence on population change. The relative influence of weather and climate will, of course, vary enormously in different cases and a more detailed discussion of the arguments which have arisen is left to the next section.

5.5 Evidence that weather determines population change in the field

In Chapter 2 we quoted the view of Howard & Fiske (1911) that weather acted as a catastrophic agent—killing a very variable proportion of insects from time to time. Andrewartha & Birch (1954) have also emphasized that variable weather determines population change.

Locusts, which live in semi-desert country where the amount of rain is very variable, are especially affected by weather (Uvarov 1931). However, their uneven distribution and their migratory behaviour make a meaningful census extremely difficult, and life-table information is hard to obtain, even for part of a generation. Key (1945) sought an explanation for the fact that the Australian plague Locust *Chortoicetes terminifera* produced swarms in some years but not in others. Without detailed population estimates, life tables, or physiological studies, the methods available to him at that time were limited. First Key investigated the possibility that there was a correlation with the 11-year sunspot cycle; no such relationship could be found. Next, it was known that the locusts needed damp, bare earth for oviposition and favoured tussocky grasses for food and shelter; the locusts were particularly numerous in the south-east quarter of Australia, in the climatic zone where rainfall was light and very variable. Believing that rainfall must influence the growth of the grasses on which the locusts fed, it was natural to see if there were any special features in the weather records in those years when swarms were reported. Key tested for a correlation between rainfall and the years of swarming. Again no clear relationship emerged. After trying many different combinations of weather measurements, Key found the best correlation with a 'rainfall index' which was influenced by the monthly distribution of rain. The index was obtained by multiplying the total number of inches of rain from October to February by the number of months in the same period in which more than 2 in. of rain were recorded (zero being counted as one half) (Fig. 5.6). With a rainfall index less than 16, no swarms were observed. Ten of the years gave rainfall indices between 17 and 69, and in six of these swarms were recorded. There is no valid way of testing the significance of such a result because it was selected as the best of many correlations examined. Although the limits were chosen so as to get the most favourable result, no swarms were observed in 1916 when the index was 69, whereas swarms were recorded in two years with an index less

D

than 20 (1922 and 1955). The weather measurements certainly do not account for *all* the variation in population but what other factors might be involved remains unknown.

The effects of climate and weather in Africa and Arabia on the

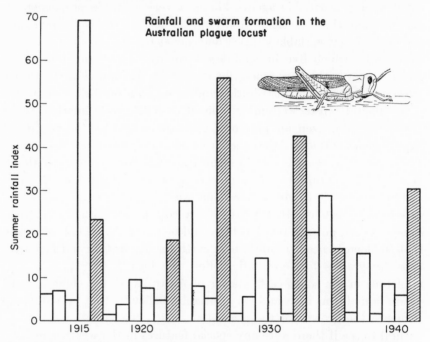

Fig. 5.6 The crosshatched histograms are for the years in which swarming of the Australian plague locust, *Chortoicetes terminifera*, took place. The rainfall index was obtained by multiplying the total number of inches of rain from October to February by the number of those months in which there was more than 2 in. of rain (zero counting as ½). The data were collected at Bogan–Macquarie, New South Wales, Australia, Key (1945).

spectacular outbreak of the desert locust *Schistocerca gregaria* in 1967 and 1968 provide a particularly dramatic story, which is well told by Baron (1972). The thousand-mile migrations of this locust are determined by wind direction, which normally blows towards the intertropical convergence zone which is a semi-permanent but mobile frontal system where rain is most likely to be falling. So by riding the wind, the locusts are carried to places where there is damp soil suitable for oviposition and also for seed germination and the development of the perennial plants which will provide the food for the locust hoppers when hatched. Rarely, the winds carry swarms out to sea, as in October 1954, to perish in the Atlantic (Johnson 1969, Fig. 19.3). The locust successfully exploits the vagaries of the weather to eke out a chancy life in the tropical semi-desert. In any one place its numbers can increase only after rains. So, as Uvarov (1931) proclaimed, there is a close causal correlation between weather and changes in locust numbers. The processes *regulating* the population level of locusts and why one locust species and not another is the common species are unexplained, and will remain so without the necessary life-table information. This will be very difficult to provide, but we must not be unduly discouraged to find that some important problems cannot yet be solved. We are only beginning to solve the simpler problems. What now seems impossible may take a little longer.

Davidson & Andrewartha (1948a, b) made an interesting attempt to interpret the change in the numbers of *Thrips imaginis*, a tiny plant-sucking insect, in roses grown in the garden of the Waite Institute at Adelaide, Australia. This pioneer study in the use of multiple regression

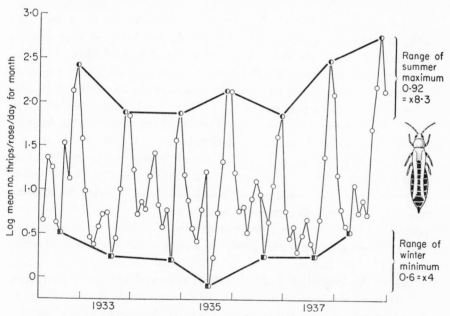

Fig. 5.7 Mean monthly population counts of adult *Thrips imaginis* in roses at Adelaide, Australia. Data from Davidson & Andrewartha (1948).

analysis has led to a lot of argument about methods and interpretations. They counted the number of *Thrips* in a sample of 20 roses every week over a period of 14 years, but they did not record the total number of roses available, and did not study the plants in which the *Thrips* bred, so the figures represent an unknown variable fraction of the real *Thrips* population. Figure 5.7 presents some of their figures. Each year the number of *Thrips* in the roses reached a peak towards the end of November. Davidson & Andrewartha assessed the effects of weather on the *Thrips* numbers by seeking correlations with meteorological data. First they converted the *Thrips* figures to logarithms and took means (with or without adjustments) over various periods near the peak which gave seven different ways of expressing the size of the peak, each of which was treated as the dependent variable Y in the multiple regression formula:

$$Y = b_0 + b_1 x_1 + b_2 x_2 + b_3 x_3 + \ldots + \ldots + b_6 x_6$$

where b terms are constants, and the x terms are various temperature and rainfall measurements for different months before the population peak.

The computation finds the best values for the constants b_0, b_1, b_2 etc., to solve the equation. If any particular value of b is very small it means that that particular x term has no significant effect on Y: it can be eliminated and the calculation is then repeated with only the more important variables. In this way Davidson & Andrewartha were able to explain up to 84 per cent of the variance in Y from the changes in four weather terms: x_2, the number of day degrees up to 31st August; x_4, day degrees in September and October; x_5, day degrees in August of the previous season; and x_3, the rainfall in September and October. This is strong evidence that most of the variation in the *Thrips* counts depends upon the weather. Although we have learned to distrust the results of multiple regression because it is always possible that such correlations are not causal, we consider that Davidson & Andrewartha made quite a good case and produced a reasonable predictive equation for population change.

Later Andrewartha & Birch (1954) reviewed this work and drew a further conclusion which seems to us unwarranted. They said (l.c., p. 582) 'not only did we fail to find a "density-dependent factor" but we also showed that there was no room for one'.

The method of numerical analysis used by Davidson & Andrewartha was designed to discover the cause of population change, and would not *directly* reveal the presence of a density dependent factor, because it was not designed to do this. The idea that there was 'no room' for a density dependent factor suggests that the factors causing population change and any density dependent factor should have an additive effect. If, as we believe is the case, the two oppose each other, then their conclusion is unsound; it has been criticized by Kuenen (1958) and Smith (1961).

It is illuminating to test in a population model how the presence of density dependent mortality affects the correlation between the observed numbers and a mortality which is represented by a random variable. Suppose we model the population changes of an insect with one generation a year which has a fixed 10-fold rate of increase. If

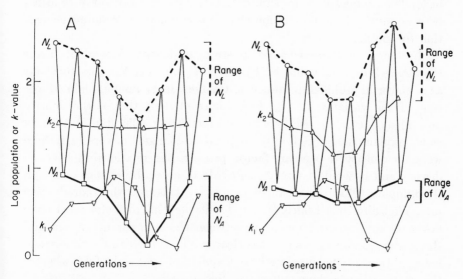

Fig. 5.8 In models A and B an adult population N_A is given given a 10-fold reproductive rate: then a random key-factor which is identical in each model acts to leave a population density of larvae, N_L which is plotted on the top line.

Model A has a weak density dependent factor

$$k_2 = 1 \cdot 275 + 0 \cdot 1 \log N_L$$

k_1 explains only 32 per cent of the variance of $\log N_L$.

Model B has a strong density dependent factor

$$k_2 = -0 \cdot 075 + 0 \cdot 7 \log N_L$$

k_1 now explains 91 per cent of the variance of $\log N_L$.

populations of larvae and adults of generation n, L_n and N_n, are represented by their logarithms, the annual increase is $1 \cdot 0$. Let us represent the catastrophic effects of weather by a random mortality factor k_1 (Fig. 5.8) which kills a variable proportion of the newly hatched larvae. Suppose that the only other mortality factor k_2 is density dependent. The question we ask is: does the change in k_1 explain the variance in the larval population (N_L) better in the presence of a weak or a strong density dependent factor?

Model A includes a weak density dependent factor,

$$k_2 = 1 \cdot 275 + 0 \cdot 1 \log N_L$$

which is unable to eliminate trends caused by the random changes in k_1. The changes in k_1 explain only 29 per cent of the variance in N_A because much of the population change is cumulative from one generation to another.

In Model B there is a much stronger density dependent factor represented by the value of $k_2 = -0 \cdot 075 + 0 \cdot 7 \log N_L$ and this acts as a powerful stabilizing influence. Changes in k_1 now explain 91 per cent of the variance in N_L. It becomes obvious that if the density dependent factor k_2 was exactly compensating (Chapter 2) then k_1 would explain 100 per cent of the population change. Conversely, if there were no density dependent factor present, the relationship between density and k_1 would be very small indeed, unless by chance or design (Reddingius 1971) the mean of k_1 exactly balanced the reproductive rate. We claim that a fairly strong density dependent factor must have acted on the *Thrips* population between the summer maximum and the winter minimum. Its effect is clear in Fig. 5.7 where the differences between the summer maxima have a spread of over eight-fold, whereas the winter minima show only a four-fold scatter.

A more direct test for density dependence in the mortality during the period of population recession is to plot the percentage change from the maximum to the next minimum, or the corresponding k-value, against the peak population. High population peaks are followed by high k-values but the points do not fall tidily on a line, which shows that the density dependency is accompanied and partly obscured by some random changes.

Andrewartha & Birch (1954) argued that weather factors determined population change and this has sometimes been restated loosely 'weather controls the population'. Nicholson (1933, 1954) took an uncompromising stand on this issue, arguing that the various ways in which weather acted on the physiology of insects could not be density dependent. Therefore weather could not control the population. We take the view that both these arguments are basically correct and do not really contradict each other. The apparent contradiction comes from two very different meanings of the word 'control' (Varley 1963).

'Control' in Nicholson's sense is a stabilizing density dependent process, like that implied by negative feed-back in cybernetics. For this process we have used the word 'regulation' following Kuenen (1958), Klomp (1962) and Bakker (1964). We agree that weather factors cannot act in this way. Arguments that weather can be density dependent usually include the idea of a limited favourable niche in which a small number of animals survive, whereas those outside the niche suffer lethal weather conditions. As the limit is set by the size of the niche we prefer to regard this as limitation by space rather than limitation by weather.

In the model B in Fig. 5.8, control in the Nicholson sense is the result of the density dependent factor k_2 which *regulates the population*.

The kind of test used to prove that climatic factors 'control' populations would show for the model in Fig. 5.8B that k_1 was *determining population change*.

The two components of the mortality in the model, k_1 and k_2 are both important, but their properties and effects are quite different. In the literature of economic entomology the word 'control' is used indiscriminately for these two processes which we distinguish by using the words determination and regulation. This is not the place to review all the confusing uses of the word 'control' in recent publications but it can often be replaced by words like 'kill', 'mortality' or even 'an insecticide application' without change of meaning. Wherever possible we have avoided the word. Where readers find the word in this book or elsewhere they should pause and consider carefully what kind of meaning is to be attached to it.

CHAPTER 6

LIFE TABLES AND THEIR USE IN POPULATION MODELS

6.1 Synopsis

If we devise suitable census methods and count the numbers in the stages of an insect which has discrete generations with an annual life cycle, the results can be expressed in a variety of ways. The numbers can be simply converted into percentages of the original number of eggs, but this underestimates the relative importance of mortalities acting towards the end of the life cycle.

Using a few simple rules the results can be presented as a *Life Table* in which different mortality factors act in sequence on the successive developmental stages. If parasites or predators are an important cause of death, then life tables should be prepared for them also. From a parasite life table we estimate the parasite's searching efficiency—the area of discovery.

When we put all that we have measured into a simple population model we can check if the population is stable and if so the level at which the population stabilizes. Changes in the constants put into the model will change the outcome. From as little as one complete life table we can begin to find how sensitive the level of population balance is to different mortality factors, which may affect parasite and host in different ways. Additional mortality can either increase or decrease the level of balance, depending on the way it affects the factors which regulate the population.

6.2 Introduction

When we have counts within a generation of the changing numbers of an insect species in the field there are advantages in expressing the results in the form of a life table. The aim is to show the effects of

94

successive events in their natural order and to express their numerical effects in various convenient ways.

In human demography the life table describes the survival of an imaginary cohort of individuals in relation to their age. This presentation might be suited to insects with overlapping generations, but has been little used because an adequate census of insects with overlapping generations is very laborious.

With insects which have a single generation in the year, the census is simplified because age specific mortality is also seasonal. The census on which life tables are to be based must count the living insects in samples of known size and should ideally count the numbers which die in each time interval and identify the causes of death.

The biology of the insect limits the type of census which can be used and so determines the type of information most readily obtained, and this affects the way in which the life tables are constructed. Different census methods may have to be used for the egg, larval, pupal and adult stages. Indirect methods of counting by means of attractants or baited traps may provide a population index for a particular stage; capture, mark, release and recapture methods give an indication both of population size and the survival rate of the active stage studied. This must be supplemented by details for other stages of development before a life table can be constructed. Two basically different sampling methods have proved specially useful, and they are discussed in this and the next chapter.

6.3   Accumulative trapping

The traps use the insect's own activity on a journey which each individual insect normally undertakes only once in its life history. The traps are set over a period of a few weeks so that all the insects, or a known proportion of them, can be accumulated and counted from a known area as they reach the critical stage of development. The sum of the numbers entering the traps over the period is used directly to estimate the population density, and the figure is entered in the life table. Three types of trap were used for the winter moth life table (see Chapter 7). They counted adult winter moths, fully grown larvae and adult parasites.

6.4   A sequence of successive samples

Counts of the insects in a series of equivalent samples taken over its life cycle are often the most convenient method for very abundant insects, if accumulative trapping is impracticable. Reference to Fig. 1.1 will remind you of the difference between a cumulative curve whose properties are exploited by accumulative trapping, and a population curve which is sampled at an instant of time. The sample may be from such units as a measured length or weight of a tree branch as used by Morris (1963) in his study of the spruce budworm, which is examined further in Chapter 8. Alternatively the sample may be from a quadrat or other defined area. The results of samples are more informative and their analysis is simplified if the insects leave some durable record of their presence as do bark beetles, leaf miners and gall insects. When a predator or parasite kills a larva in a gall it may perhaps remain in the gall long enough to be counted, or specific traces of its former presence may remain in the gall after it has gone.

The knapweed gall-fly (*Urophora jaceana*, Diptera, Trypetidae) is abundant in many parts of Europe and is especially easy to study in autumn and winter. At this time the dead flower-heads of the knap-weed, *Centaurea nemoralis*—(Compositae), contain galls of different sizes. Some galls contain only a single chamber or cell, and others may contain 10 cells or more, each of which must originally have contained a larva of the gall-fly. By the end of the summer and through the following winter some gall cells may also contain one or more of the different parasites and predators. Table 6.1 lists the contents of some galls and separates them into different categories, which are illustrated in Fig. 6.1 (see also Varley 1947).

We can treat such figures in many different ways. In Table 6.2 column 3 the number found in each category is expressed as a percentage of the total.* Because this does not take into account the order in which events took place during the summer, this presentation is unhelpful and indeed misleading. The figures might be expressed as numbers per square metre; this enables us to make comparisons with samples taken at other times of the year. To assess the importance of different types of mortality, however, we must arrange the biological events in their correct sequence, and for this we need detailed knowledge of the life histories and biology of the insects concerned.

6.5   The life history of the knapweed gall-fly

The gall-flies are active early in July and the female fly selects a small

---

* The figures are given to the *nearest* tenth of a percent; the total so calculated is therefore unlikely to be *exactly* 100·0.

(a) *Eucosma hohenwartiana*, which leaves the plant to pupate in the ground.
(b) *Metzneria metzneriella*, which passes the winter as a full grown caterpillar in a cocoon in the flower head.
(c) *Euxanthis straminea*, whose presence can be recognized by a hole it makes in the plant stem just under the flower head; this widespread species did not appear in this particular census.
7   The healthy larvae of the gall-fly *Urophora* normally go into winter diapause and do not pupate until the late spring.

6.6   The life table

In constructing a life table from figures like those in Table 6.1 a number of rules must be followed:

Rule 1. Where events such as death from definable causes are reasonably well separated in time, they are treated as if they are entirely separated with no overlap.

Rule 2. Where events overlap seriously in time, it may be convenient to consider them as if they acted exactly contemporaneously which we do with items 4 of Table 6.1.

Rule 3. Any animal discovered must be considered either as alive and healthy or as certain to die or already dead from some cause. If a larva is already parasitized and host and parasite are alive, the host is counted as certain to die. The parasite must be recorded as the cause of death.

Rule 4. No animal can be killed more than once. If the host is successively attacked by two parasites, the death of the host must be credited to the first parasite. If the second parasite is, in fact, the eventual victor it must be credited with the death of the first parasite. The second attack is entered in the life table of the parasite, but not that of the host.

Only in this rigorous but somewhat arbitrary manner can the accounts be balanced.

When life tables for both the host and the parasites are constructed, some other firm assumptions must be made:

Rule 5. Destruction of the contents of galls must be assumed to be random and to affect host and parasite equally unless there is contrary evidence. If this assumption is not accurate, or if sampling errors affect the figures, then small or even considerable allowances must be made during the process of constructing life tables.

flowerbud of *Centaurea nemoralis* and inserts a small group of eggs into the space between the florets and the bracts. The newly hatched larva enters a floret and bores down into the ovary which develops to form the gall. If two or more florets are galled they usually fuse into single structure with separate flask-shaped chambers (Fig. 6.1) each containing a larva. By the end of the summer the gall is woody and the plump larvae remain head downwards with the black sclerotized posterior end plugging the exit to the gall. In the following June, such larvae turn round to face the opening into the gall cell and undergo metamorphosis; the colourless part of the larval cuticle turns brown and hardens to form a puparium, within which there is a complex process of pupation. Eventually an adult fly develops, emerges from the gall, stretches its wings and is ready to mate. The females then lay their eggs.

Census studies through the summer show that a succession of parasites and predators attack the early stages of the gall-fly. In Table 6.1 different categories of gall contents have already been arranged in sequence:

1   The total number of gall-cells examined is put first because the galls must be formed by the plant before anything can happen to them.
2   Some of the gall-cells were found to be empty with no sign of larval feeding. The larva must have died at a very early stage of growth.
3   Dissection of the gall-fly larvae during the summer showed that the Chalcid parasite *Eurytoma tibialis* puts its egg into the very small larva of the host. When the host gall-fly larva is fully grown, the *Eurytoma* larva (still very small) causes premature pupation in the host which forms a hard brown puparium whose contents the *Eurytoma* larva soon entirely eats.
4   Included in this item are four different kinds of parasite larvae. They are easily recognizable, and, with one exception, only one parasite individual can mature on a fully grown gall-fly larva (see Fig. 6.1). The exception is *Tetrastichus* sp. B. which is a gregarious internal parasite.
5   A small bright orange midge larva *Lestodiplosis* (a Cecidomyid fly) is sometimes found in a gall cell with a dead gall-fly larva which it has killed and partly eaten.
6   Gall-cells are frequently entered by caterpillars of three species of moth. These can complete their development in ungalled flower heads, but, if they find a gall, they eat its fleshy parts as well as any contained insect larvae. When the caterpillars have finished feeding, the gall cell contains nothing but frass and the previous history of its occupants cannot be directly reconstructed. The common caterpillars are:

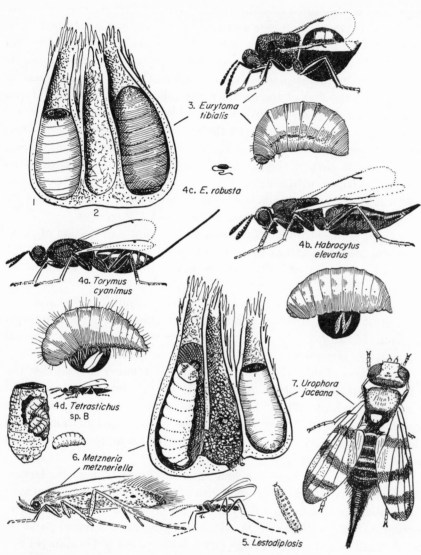

Fig. 6.1 Contents of knapweed galls listed in Table 6.1.

1 Larva of gall-fly *Urophora jaceana* rests head downward.

2 Gall cell empty—gall-fly larva died when very small.

3 Gall-cell contains brown puparium of gall-fly, now head upward. Inside is the larva of *Eurytoma tibialis* (Rt) from which the black adult *Eurytoma* (above rt) would develop by July.

4a *Torymus cyanimus* lays smooth eggs. The larva is active and hairy, here shown beside remains of host, on which are two egg shells.

Table 6.1 Knapweed gall census September–October 19 Contents of gall cells in 10 m² quadrats. The percentage colu is related to the grand total of 906.

			Number	Percenta
1	Gall cells examined		906	
2	Empty cells—gall tissue not eaten		19	2·1
3	Larva of Chalcid parasite *Eurytoma tibialis* in host puparium		349	38·6
4	Miscellaneous Chalcid parasites			
	Torymus cyanimus	21		
	Habrocytus elevatus	46		
	Eurytoma robusta	35		
	Tetrastichus B	7		
			109	12·0
5	Predatory midge larva *Lestodiplosis*		3	0·3
6	Gall cells destroyed by caterpillars of			
	Eucosma hohenwartiana		103	11·4
	Metzneria metzneriella		50	5·5
7	Healthy larvae of gall-fly *Urophora jaceana*		273	30·2
	Totals		906	100·1

4b *Habrocytus elevatus* lays groups of eggs which have a rough s The larva is shiny, almost hairless and inactive.

4c *Eurytoma robusta* adult and larva very like *E. tibialis*, but the is black and the larva feeds externally.

4d The gregarious larvae of *Tetrastichus* sp. B are found in a dead larval skin.

5 *Lestodiplosis* is a small midge, whose predatory larva is br orange in colour.

6 A caterpillar of *Metzneria* has destroyed the contents of two cells, and has formed a silken cocoon in one cell.

7 A healthy gall-fly larva and an adult female *Urophora jaceana*.

The justification for this complex process is that it condenses and summarizes the figures. The raw census figures for the gall-fly fill 250 foolscap pages. The life tables reduce this to a hundredth. The census figures alone do not reveal the biological processes that have operated and may mean remarkably little until interpreted. When suitably interpreted, the life table figures can be used for the construction of population models. In constructing the complete life table for the knapweed gall-fly, 4 periods of the life history need separate consideration.

Period A   *July*—Egg to gall formation.
Period B   *Summer and Autumn*—Gall formation to the onset of winter.
Period C   *Winter*—Hibernation (diapause).
Period D   *Spring and early Summer*—when various parasites, predators
and weather cause further changes before the flies emerge as adults.

Accuracy in this type of census is inversely related to rates of change. The most accurate census is before the winter, at the end of Period B. Figures like those in Table 6.1 enable us to reconstruct the events in this period. The examination of galls in either Period B or C forms a good class exercise in which students can do the counting and assess the causes of death for themselves.

We know that the events took place in the order in which they have been arranged in Table 6.1.

1   The total number of gall-cells found in the 10 m^2 represents the number of larvae which formed galls.

2   The number observed which died without feeding is 19. However, the fraction which died at this time is not $19/906 = 2 \cdot 1$ per cent because this event was not observable in the 153 cells damaged by caterpillars. The true fraction is

$$\frac{19}{906-153} = 2 \cdot 5 \text{ per cent}$$

Assuming Rule 5 applies, the real number dying at this stage in the 906 cells must have been 23. The caterpillars must have destroyed the evidence in about four cases in the 10 m^2.

3   *Eurytoma tibialis* lays eggs in the very small larvae of the gall-fly. If the host is subsequently destroyed by either caterpillars or by some ectoparasite, direct evidence of parasitization is lost. So the true percentage of parasitism by *Eurytoma tibialis* is

$$\frac{349}{349+273} = 56 \text{ per cent}$$

and the estimated number originally present in the 883 growing larvae of the host is 495, which is quite different from the number 349 observed. All trace of some 146 *Eurytoma tibialis* must have been obliterated by the effects of other parasites and caterpillars in sections 4, 5 and 6 of Table 6.1.

4 and 5   The miscellaneous parasites and the larvae of the midge *Lestodiplosis* attack fully grown gall-fly larvae and so act contemporaneously (Rule 2). The total number of cells containing these species was 112 and the presence or absence of successful attacks by these species was verifiable in 734 cells. Hence the percentage of gall-fly larvae destroyed was 15 per cent, which is entered in column 2 of Table 6.2. The number killed by these species, 58, is entered in column 5. It is far less than the total number of cells found containing these species, 112, because of Rule 4. We estimate that about half the gall-fly larvae attacked by these other parasites were already certain to die of some other cause. These miscellaneous parasites must be credited with

Table 6.2   Knapweed gall-fly life table for Period B based on Table 6.1 for 10 m^2.

Column 1	2	3	4	5	6	7	8
	Fraction observed killed	% killed	Fraction survived	No. killed	No. alive	k-value	Log survive per m^2
Initial number o Gall Cells					906		1·957
Larvae died young	$\frac{19}{753}$	2·5	0·975	23	883	0·011	1·946
Larvae parasitized by *Eurytoma tibialis*	$\frac{349}{622}$	56	0·44	496	387	0·358	1·588
By miscellaneous parasites and *Lestodiplosis*	$\frac{112}{734}$	15	0·85	58	329	0·070	1·518
Killed by caterpillars	$\frac{153}{906}$	17	0·83	56	273	0·082	1·436

the death of many *Eurytoma tibialis* and the proportion of records are transferred to the life table of that species.

6   The three species of caterpillar commonly feed in the flower heads in summer and eat the fruits and usually destroy the contents of galls they encounter. Here 153 cells were found damaged and such damage was never obscured by other events, so the proportion of gall-fly larvae destroyed is taken to be 153/906 = 17 per cent, leaving 273 survivors.

7   The number of healthy gall-fly larvae recorded in this sample from 10 m² agrees exactly with this figure.

6.7   Correcting and completing the life table

The method shown in Table 6.2 to derive a life table for Period B is satisfactory for a homogenous sample, but the quadrats sampled in October 1935 included only the flower heads on the standing stems, and did not include galls which might have been detached from the stems. In fact, by the end of the summer many flower heads had already been nipped off by wood mice. The mean number of gall-cells found per square metre in standing flower heads in August was 148. The October sample in Table 6.1 and 6.2 had the mean already reduced to 91. By May and June of 1936, there was a further reduction to 57 gall-cells per m². Quadrats had been carefully searched for fallen galls, and many of the galls found on the ground had been opened by mice and the contents extracted.

The basic figure for Section B of the corrected life table in Table 6.3 is the initial number of gall-cells in the last line of Section A. The estimates for the *k*-values and for the *percentage* mortality from different causes in Table 6.2 are valid and have been copied into Table 6.3 and are set in bold type. The other figures in Section B can be derived at once from them. If you check them, you will find that because the percentages of mortality have been given only to two significant figures and the *k*-values to three places of decimals the two calculations do not give precisely the same results.

Period A

Because the duration of the egg stage is less than the total period of oviposition, no census count of eggs represents the number of eggs laid per m² in the season. The number of eggs laid can be obtained only indirectly from the fraction of eggs surviving to form galls and from the

Table 6.3 Corrected life table for the knapweed gall-fly.

Column 1	2 % killed	3 No. killed per m²	4 Alive per m²	5 k-value	6 Log no. alive per m²
Period A					
Adults emerged were 0·425 ♀			6·9		0·839
71 eggs/♀			209		2·320
Infertile eggs	**9**	19	190	0·040	2·280
Larvae fail to form galls	**22**	42	**148**	0·110	2·170
Period B					
Larvae die young	**2·5**	3·7	144·3	**0·011**	2·159
Parasitized by *Eurytoma tibialis*	56	81	63·2	**0·358**	1·801
Miscellaneous parasites	15	9·5	53·8	**0·070**	1·731
Killed by caterpillars	17	9·3	44·5	**0·082**	1·649
Period C					
Winter disappearance	**61·5**	27·4	17·1	**0·415**	1·234
Killed by mice	**64**	10·0	6·1	**0·444**	0·790
Period D					
Killed by birds	3·8	**0·2**	5·9	0·017	0·773
Larvae dead	27·5	**1·6**	4·3	0·139	0·634
Parasitized by *Habrocytus* etc.	16	0·7	**3·6**	0·078	0·556
Pupae drowned	44·5	1·6	**2·0**	0·255	0·301

148 gall-cells found per m². Egg mortality in 1935 appeared random within batches of eggs and amount to 9 per cent ($k = 0·04$). Comparing the frequency distribution of eggs with that of gall-cells suggested that if this mortality of eggs and young larvae was random the change in the frequency distribution represented 29 per cent mortality ($k = 0·15$).

Because these act successively we can enter k-values in column 5 of 0·04 for infertility and $0·15 - 0·04 = 0·11$ for the larvae failing to form galls. Adding these k-values to the log of 148 (2·17) we obtain a logarithm of 2·32, whose antilog, 209, is the number of eggs per m².

These eggs were laid by an estimated number of 6·9 adults per m². The proportion of females in a large sample was 0·425, from which we estimate the number of eggs laid per female as $209/(6·9 \times 0·425) = 71$.

An alternative way of showing these figures, which is more suitable if either the proportion of females or the egg production per female is variable, is to represent the proportion of females by its equivalent k-value. The log of 0·425 is $\bar{1}·628$, so the k-value is 0·372 (Chapter 1). Individual females sometimes laid over 200 eggs. If some arbitrary figure for the potential egg production per female is adopted, which is slightly greater than any observed mean value, then any reduction in mean egg production can be expressed as a k-value. If we take a figure of 200 for the potential egg production per female, then the difference between this and that which the flies appeared to lay (71) can be expressed by $k = 0·45$.

Period C

Mice and stormy weather progressively reduced the number of galls on the standing stems of knapweed during the winter. Careful search for galls on the ground as well as on the stems in May and June 1936 gave a mean value of 57 gall-cells per m². This loss of 91 gall-cells per m² from the 148 found in Period B is termed 'winter disappearance'. This is 61·5 per cent loss, or $k = 0·415$. Of those galls found on the ground 64 per cent of the cells had been opened from the base by mice and the contents abstracted ($k = 0·444$). Putting these k-values into column 5 of Table 6.3 we calculate that the log of the population of gall-fly larvae was 0·79, which represents a population density of 6·1 per m².

Period D

In the early summer, the surviving gall-fly larvae, many now in galls which had fallen to the ground, pupate and eventually emerge as flies, leaving an empty puparium in the gall-cell as evidence of successful emergence. But throughout this period various Chalcid parasites like *Habrocytus elevatus*, *Macroneura vesicularis* and others attacked and laid eggs on the gall-fly and parasite larva in the few galls still on the standing stems. Then in July heavy rain flooded the census area. When the water receded many larvae and pupae in fallen galls were evidently dead. The bold figures in column 3 are an approximate representation of the numbers found which varied from sample to sample. The figures in the other columns are all derived by calculation from the figure of 5·9 in column 4 of Period C.

6.8 Life tables for the parasites

Varley (1947) used the census figures to derive life tables for the four major parasites. The methods are identical with those just used for the life table of the gall-fly, and we will take the life table for *Eurytoma tibialis*

Table 6.4 Corrected life tables *Eurytoma tibialis,* a parasite of the knapweed gall-fly.

Column 1	2	3	4	5	6
	% killed	No. killed per m²	Alive per m²	k-value	Log no. alive per m²
Period B					
Adults emerged			2		
0·52 females 78 eggs/♀			**81**		1·908
Miscellaneous parasitism	**15**	12	69	**0·070**	1·838
Killed by caterpillars	**17**	12	57	**0·082**	1·756
Period C					
Winter disappearance	**61·5**	35	22	**0·415**	1·341
Killed by mice	**64**	14·1	7·9	**0·444**	0·897
Adjustment Missing	42	3·3	4·6	0·235	0·662
Period D					
Killed by birds		0·4			
Misc. parasitism	24	0·7	3·5	0·118	0·544
Drowned in floods	53	1·84	1·66	0·324	0·220
Adults emerged			1·66		

as an example (Table 6.4). The number of adults emerging in each generation was estimated (with some uncertainty) from the numbers of live pupae and pupal exuviae from which adults had recently emerged. In Table 6.3 the gall-fly larvae recorded parasitized by *Eurytoma tibialis* were 81 per m^2, so this figure has been transferred to column 4 of Table 6.4. The number of eggs laid per female was calculated from the first two figures in column 4 and from the proportion of females. The attack by other parasites and by caterpillars, winter disappearance and damage by mice seemed likely to be indiscriminate. We therefore assume that these factors killed gall-fly larvae whether already parasitized or not and the bold figures in column 5 of Table 6.4 have been transferred here from the gall-fly life table; from them, all the other figures in Periods B and C can be derived, leaving an expected number of *Eurytoma* alive after the winter as 7·9 per m^2. The early spring census recorded only 4·6 live *Eurytoma* per m^2 left. These figures do not differ significantly because they are usually means for 10 m^2 samples with a high variance. But it is better to make a firm adjustment by introducing the term 'missing' into the life table to cover what may be either a real difference in the incidence of some of the mortality factores, or a sampling error.

6.9   Inferences from life table figures

If you run your eye down column 5 of the life tables for the gall-fly and its parasite the biggest values appearing in both are in the winter Period C, where the two k-values add up to 0·859. The k-values attached to predation by caterpillars and by birds are also fairly large, so that the total sum of the k-values of indiscriminate mortality affecting both the gall-fly and its parasite amount to 1·1. This is equivalent to 92 per cent mortality and only 8 per cent survival. The number drowned in floods is omitted as being unlikely to recur.

Looking at the life table for the gall-fly, the effect of *Eurytoma tibialis* appears small when compared to the effect of this indiscriminate mortality. Varley (1947) used simple population models to see how indiscriminate mortality might affect the population balance between host and parasite. First the area of discovery of the parasite was estimated from the formula

$$a = 2 \cdot 3 \frac{k_p}{P}$$

(re-arrangement of formula (4.4)) where k_p is the k-value of parasitism on the host and P is the parasite population density. Using the values for k_p in Table 6.3 and of P in Table 6.4, the area of discovery of *Eurytoma tibialis* is estimated at $0.4 \, \text{m}^2$. A second, perhaps less accurate, value for the area of discovery was estimated for the following year as $0.2 \, \text{m}^2$. These estimates do not differ significantly and for the calculations which follow it will be assumed that for a parasite population of 2 per m^2, the area of discovery is $0.3 \, \text{m}^2$.

If we wish to speculate about the influence of indiscriminate mortality upon the interaction between the parasite and its host we need to make firm but plausible assumptions about parasite behaviour. Varley (1947) used Nicholson's theory with its assumption of a fixed area of discovery in a population model for the gall-fly. In Chapter 4 we saw that mutual interference between parasites could cause the area of discovery to decrease as parasite density increased. Figure 6.2A shows the expected values of the area of discovery and of the killing power of the parasite population for different values of the mutual interference constant m. When $m = 0$ we have the Nicholson assumption, and with values of $m = 0.3$ and 0.5 we cover a range of values similar to those which have been observed with other parasites. The values of the quest constant Q in Fig. 6.2A have been calculated so that for each assumed value of m the area of discovery equals the observed value of 0.3 when the parasite population is 2.

To find how the value of the mutual interference constant affects the outcome, we use a population model and calculate the 'steady' densities of host and parasite when the host increase is $\times 18$ in each generation and the indiscriminate mortality of hosts surviving parasitism and of the parasite larvae are both 90 per cent. Host adult density N_A is always the same so after reproduction the host density will be $18N_A$ and the density before the action of the 90 per cent indiscriminate mortality must be $10N_A$. This can be represented by the following model:

$$\text{host adults } N_A \xrightarrow{\text{reproduction}} 18N_A \xrightarrow{\text{parasitism}} 10N_A \xrightarrow[\text{mortality}]{\text{indiscriminate}} N_A \text{ adults.}$$

Similarly, parasite adult densities P are always the same and after reproduction parasite density $= 18N_A - 10N_A$ (the number of hosts destroyed). Also if the parasites suffer 90 per cent indiscriminate mortality, parasite density before this must be $10P$. The model for this is:

$$\text{parasite adults } P\xrightarrow{\text{reproduction}} 8N_A = 10P \xrightarrow[\text{mortality}]{\text{indiscriminate}} P \text{ adults.}$$

Thus we have to calculate the conditions under which $8N_A = 10P$.

Table 6.5 shows the steady densities calculated using formula (4.12).

$$k_p = \frac{QP^{1-m}}{2\cdot3}$$

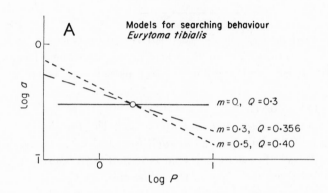

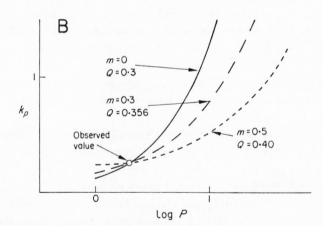

Fig. 6.2

A Models for the searching behaviour of *Eurytoma tibialis*.

B Their effects on the host population.

The point where the curves cross represents the observed value of the area of discovery of *Eurytoma*, which was 0·3 (log = $\bar{1}$·477) at a parasite population of 2 (log = 0·3).

Table 6.5 Population models for the knapweed gall-fly.

Assumed parasite properties		Assumed indiscriminate mortality	Calculated steady density		Assumed indiscriminate mortality	Calculated steady density	
m	Q		*Eurytoma*	Gall-fly		*Eurytoma*	Gall-fly
0·0	0·30	90%	2·0	2·5	0%	9·7	0·57
0·30	0·356	90%	2·1	2·6	0%	20	1·2
0·50	0·40	90%	2·2	2·7	0%	52	3·1

which is solved for P using the assumed values for Q and m and a value of $k_p = \log 1 \cdot 8 = 0 \cdot 256$, because k_p is the difference between the logarithms of the initial and final population after parasitism, which are in the ratio 18 : 10.

When the indiscriminate mortality is 90 per cent the calculated steady density of *Eurytoma* and of its host the knapweed gall-fly is between 2 and 3 adults per m² in each case and there would be 45–50 gall cells per m². These figures compare well with census figures. We can speculate about the effect of the indiscriminate mortality by finding the steady density in a model in which the indiscriminate mortality is given a zero value, as in the right hand part of Table 6.5. If the population is to be stable the parasite must now kill 17/18 of the gall-fly larvae so that $k_p = \log 18 = 1 \cdot 256$ in equation (4.12) so the parasite population must be higher than when $k_p = 0 \cdot 256$.

The way in which k_p varies with parasite population is shown in Fig. 6.2B and values of P can either be read off the curves or calculated by solving the equation (4.12).

The elimination of indiscriminate mortality leads to a new steady population density in which parasite populations are considerably raised; in comparison host densities are changed remarkably little. With a mutual interference constant of 0·3 *removal* of the indiscriminate mortality *reduces* the gall-fly population from 2·6 to 1·2; with $m = 0 \cdot 5$ it rises somewhat from 2·6 to 3·1. Considering that we are estimating the effect of a 90 per cent mortality acting year after year the change in the steady density of the gall-fly is surprisingly small.

The reason why the effect is so small is not that the population is stabilized by a density dependent factor but because the indiscriminate mortality has two antagonistic effects: by reducing the effective natural increase of the host it tends to reduce its equilibrium population density, but by reducing the survival of the parasite it tends to increase the population density of both parasite and host. The net outcome is a big change in parasite density and a very small one in host density.

6.10 Discussion

A life table presents a simple summary of population change within one generation and can reduce a very large body of census figures to a form which can be readily analysed. If life tables can also be prepared for parasites then population models can be used to see how other measured factors interact. The minimum essential information is an estimate of the population density of searching parasites and of the percentage of parasitism that they cause to one host generation.

If the insect studied were a pest the likely effects of an insecticide programme could similarly be investigated. The effect of indiscriminate mortality might be very like that of an insecticide which killed equal proportions of the pest and its natural enemy. If an insecticide killed more than 17/18 of the pest in this model, then both pest and parasite would eventually be eliminated. The snag is that insecticide programmes are normally timed very accurately to fit some particular target pest and may fit other potential pests far less well. The insecticide may eliminate the effective natural enemy of a potential pest, leading to an unexpected outbreak. In current jargon these are termed 'side effects' but with simple life table data for a generation or two their prediction should be possible and they should be avoidable, at least in simple cases.

The processes and interactions even between two species are too complex to follow without a population model. This chapter has introduced modelling in its simplest form. For overlapping generations the models would be more complex and we have not yet a realistic model for predators. But the models we have used in this chapter give considerable insight into population interactions, without requiring any more complex facilities for computation than a slide rule or log tables.

CHAPTER 7

INTERPRETATION OF WINTER MOTH
LIFE TABLES

7.1 Synopsis

From life tables for the winter moth we can identify the main cause of population change from year to year, which we term the key factor. Density dependent mortality serves to regulate the population density and keeps it within limits. Such density dependent mortality may either act directly—like food limitation—or may arise through the behavioural responses of parasites and predators to their own and to their host's population densities.

Population models which represent these effects realistically enable us to see how the various components of the system interact with each other. By changing the various components in the model we can see how changed conditions in the field might affect the population.

7.2 Introduction

In this chapter we outline the ways of analysing life table figures for the winter moth. This has proved an easy insect to study because it is very abundant and has an annual life-cycle with each stage concentrated at a different time of year. The methods will be applicable to other animals with a restricted breeding season, and have indeed been adapted by workers to interpret census figures for birds such as partridge (Blank *et al.* 1967), owls (Southern 1970), tits (Krebs 1970a, b) and grouse (Watson 1971).

7.3 Winter moth life history and the census

Larvae of the winter moth are able to feed on a wide range of trees and shrubs, but they are especially abundant on oaks (*Quercus robur*), which they sometimes defoliate. Near the northern boundary of

Wytham Wood, Berks, is a fairly large area where oak is the pre-
dominant large tree. There we selected for study a place where there
were over 20 large oaks but no other major trees. Outside the study area
were dense thickets of blackthorn, quite a lot of hazel and a few birches,
some hawthorn and sallow, on all of which winter moth can feed. But

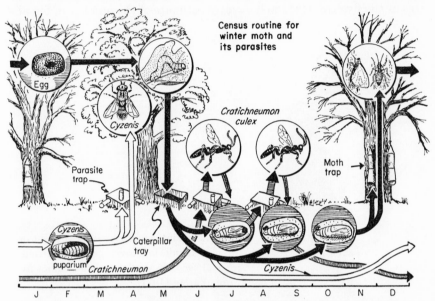

Fig. 7.1   Adult female winter moth are counted in moth traps on the
tree trunks, their larvae are counted in the caterpillar trays into which
they fall when prepupal. Larvae of the parasite *Cyzenis* are counted by
dissection of the fallen caterpillars. Adults of *Cyzenis* and of *Cratichneumon* are counted on emergence from the soil into the parasite
traps. From Varley (1971).

oaks provided the major canopy. From 1950 we restricted the census
to the insects on or under five trees, although some general observations
were also made on other trees and bushes. The seasonal census routine
and the life history of the winter moth and two of its main parasites
are illustrated diagramatically in Fig. 7.1.

Winter moth adults emerge from the soil under oak trees in November
and December. At dusk the flightless females walk to the trees which
they climb. The winged males, which rest by day in the litter, fly
actively at dusk and congregate on the lower part of tree trunks. Here
they mate with the females (Fig. 7.2) which continue to climb the trees
to lay eggs in crevices in bark and lichen high above theground.

One quarter of the females climbing each of the five trees were caught in traps like small lobster-pots made of fabric supported with wire. Two traps were placed on opposite sides of each tree, and each was arranged to obstruct one eighth of the perimeter of the tree. The total catch of females multiplied by four and divided by the total canopy area of the five trees (282 m²) gave an estimate of the number of females

Fig. 7.2 Winter moths mating at night on an oak trunk in November. The female (above) has very small wings and cannot fly. From Varley (1971).

per m² of canopy area. The number of adults per m² was twice this number because we knew there were equal numbers of males and females in the pupal stage.

Females were dissected and, though a few had as many as 300 eggs, they contained on the average about 150 eggs each. From this we estimated the number of eggs laid in each tree. We did not try to estimate the actual number laid because this would have entailed the destruction of the trees.

When the oak buds are beginning to open in early April the eggs hatch and we find the first stage caterpillars feeding in the buds, where they do great damage to the tiny leaves. By the latter half of May feeding is completed and the caterpillars spin down from the trees on silken threads, burrow into the soil, spin cocoons and pupate.

We used two methods to study the caterpillars feeding on oak. We took samples of twigs from the tree tops and counted the larvae on them; this was the only way we could assess the changes in the numbers of those species which pupated amongst the leaves. For winter moth, and for many other species which pupate in the soil, cumulative trapping was feasible. We used two trays on the ground, each of 0·5 m², under each of the five trees. We allowed rain water to accumulate in the trays so that when the larvae spun down to pupate they drowned. These larvae were identified, examined for external parasites and dissected for internal parasites. From the figures we could assess the number of healthy fully grown larvae per m² and the number parasitized by each of the different kinds of parasite.

The survival of the healthy winter moth pupae was known from the number of adults trapped the following winter. The survival of the parasites to the adult stage was estimated by traps into which adult parasites emerged from the ground. For this we used the same metal trays in which we had counted the caterpillars, but now they were inverted and their edges pushed into the ground. In three of the corners of each trap were inserted glass tubes provided with celluloid cones which restricted the return of the parasites into the dark interior of the trap. Experiments showed that an adult parasite inserted into one corner of a trap was attracted by the light and was soon recaptured at one of the other windows. Each trap again covered half a square metre so that the sum for 10 traps under the five trees gives the figure for 5 m².

Table 7.1 gives as an example the trapping results for some of the important species for the year 1955–56. These have been used in the construction of the life tables for winter moth and for one of its parasites, the tachinid fly *Cyzenis* (Tables 7.2 and 7.3).

7.4 The components of the life table

The life table is in three sections: the females counted in the moth traps in late 1955 number 413 and as the surveyed canopy area of the trees was 282 m² we arrive at the figure of 5·85 females emerging per m² of which three-quarters, 4·39 per m², by-pass the traps and lay their eggs on the trees. Multiplying the number of females per m² by 150 (eggs per female) gives the number of eggs per m² as 658.

7.4.1. *Winter disappearance*

The count of fully grown larvae was only 96·4 per m² and the k-value representing this loss (k_1) is 0·84 (= 86 per cent mortality). We call this loss *winter disappearance* although it represents the loss from all causes

Table 7.1   Trap results 1955–1956.

	No. of traps	No. of m² trapped	Number captured	Number per m²
Tree trunk nets				
Female winter moth				
Nov.–Dec. 1955	10	282	(1/4 =) 413	5·85
1956	10	282	(1/4 =) 525	7·45
Ground traps				
Adult *Cyzenis*				
in April 1955	20	10	3	0·3
1956	20	10	3	0·3
Adult *Cratichneumon*				
in July 1956	10	5	13	2·6
1957	10	5	67	13·4
Caterpillar trays				
May 1966	10	5		
Healthy winter moth larvae			415	83·0
Attacked by *Cyzenis*			31	6·2
Attacked by other parasites			13	2·6
Infected by Microsporidian			23	4·6

between the count of females up to the count of fully grown larvae. We do not think that predation on the adult females by birds causes a high mortality, and experiments showed that most eggs hatched when put on bark and twigs in the tree tops. Also, there were no obvious changes in the numbers of young and old larvae counted in the twig samples, so we do not think that mortality caused by predation on the feeding caterpillars is an important component of k_1.

However, the winter disappearance differed between trees in relation to the time of budburst, and was least on those trees which opened their buds early. Larvae which hatched on trees whose buds were firmly closed mostly failed to find food there. Such larvae emigrate by spinning a length of silk on which they are blown away from the trees. The number which disappeared from the trees was closely related to the numbers of first stage larvae which were caught on sticky traps placed under and between trees.

Table 7.2   Life table for winter moth 1955–1956.

	Percentage of previous stage killed	Number killed per m²	No. alive per m²	Log no. alive per m²	k-value
Adult stage					
♀♀ climbing 1955			**4·39**		
Egg stage					
♀♀ × 150			658	2·82	
Larval stage					$0 \cdot 84 = k_1$
Full grown larvae	86·9	551·6	**96·4**	1·98	$0 \cdot 03 = k_2$
Attacked by *Cyzenis*	6·7	**6·2**	90·2	1·95	$0 \cdot 01 = k_3$
Attacked by other parasites	2·3	**2·6**	87·6	1·94	
Infected by Microsporidian	4·5	**4·6**	83·0	1·92	$0 \cdot 02 = k_4$
Pupal stage					$0 \cdot 47 = k_5$
Killed by predators	66·1	54·6	28·4	1·45	
Killed by *Cratichneumon*	46·3	**13·4**	15·0	1·18	$0 \cdot 27 = k_6$
Adult stage					
♀♀ climbing trees 1956			**7·5**		

The figures in heavy type are those actually measured. The rest of the life table is derived from these.

7.4.2   *Larval mortality*

The dissection of fully grown larvae revealed the numbers which were certain to die because they were parasitized. Although quite a number of different parasites were recorded, it is convenient to treat two of the larval parasites separately and lump the others together. The total parasitism was usually less than 30 per cent and, therefore, it made little difference to the analysis whether the different kinds of parasitism were considered as overlapping (which was true) or as operating successively. This course was followed, even though a small error was introduced, because it conveniently simplified the subsequent analysis.

7.4.3.   *Parasitism by* Cyzenis albicans

This tachinid fly (Fig. 7.1) emerges into the parasite traps in April. The females spend about four weeks feeding on sap fluxes and at flowers whilst they mature as many as 2000 minute black eggs which develop till ready to hatch and are stored in a greatly expanded uterus. The eggs are mostly laid on the leaves damaged by caterpillar feeding (Hassell 1968) and, if lucky, are swallowed whole by a winter moth caterpillar. The larvae hatch from the egg in the fore-gut of the winter moth larvae, bore through the gut wall and enter a cell of the salivary (silk) gland. They were counted at this stage in the fully grown caterpillars when these were dissected. The life table for *Cyzenis* (Table 7.3) shows the

Table 7.3   Life table for *Cyzenis* 1955–1956.

	No. alive per m²	Log no. alive per m²	k-value
Female adults emerged 1955	0·15		
Potential eggs (2000/♀)	300·0	2·48	
			1·69
Larvae in winter moth	6·2	0·79	
			1·31
Adults emerged in 1956	0·3	$\bar{1}$·48	

enormous loss before this stage. These are the ones which remained uneaten or unlaid or those which were eaten by species other than winter moth, which would either fail to hatch or be killed as small larvae.

7.4.4   *Other parasitism*

Parasitism of larvae by other tachinids and by some ichneumonids was much lower and the figures for these are lumped together in Tables 7.1 and 7.2.

7.4.5   *Parasitism by a microsporidian*

The protozoan *Plistophora operophterae* is a microsporidian distantly related to the *Plasmodium* species which cause malaria in man. Infected cells are easily seen in the salivary glands; they are opaque and filled with spores and we think that all infected larvae die before reaching the adult stage (Canning 1960). The healthy winter moth larvae form cocoons and pupate.

7.4.6 *Pupal mortality*

The density of healthy larvae was far more than twice the density of adult females which were estimated later. At first we thought that only predation was involved, but experiments to find which predators were responsible showed that parasitism was also important. In experiments where pupae were exposed to predation, some were found to be parasitized by *Cratichneumon culex*, an abundant woodland ichneumonid which, up to this time, had not been recorded from the winter moth. However, we had counted the adults of *Cratichneumon* which emerged into the ground traps and, once we knew that it was a fairly specific parasite, we could estimate retrospectively its effects on the winter moth. The few *Cratichneumon culex* which have been reared from other species of host can be neglected.

In measuring pupal mortality we only have a direct measurement of the difference between the count of healthy larvae entering the soil to pupate and the count of the adults on the tree trunks. Some of this mortality is caused by *Cratichneumon* and we take a minimum value for this based on the numbers of adult parasites emerging. The rest of the mortality we call pupal predation; this will be discussed later on.

It is convenient to consider pupal mortality as if it was caused first by predation followed by parasitism; although there must be some overlap between the two effects. Making this assumption has little effect on our assessment of the relative importance of predation and parasitism. In the absence of precise knowledge of the timing of events, the only other assumption we could have made is that parasitism takes place first and is then followed by an equal percentage predation of both healthy and parasitized pupae. In most population models these alternative assumptions produce identical effects.

If then we assume that *Cratichneumon* acts after predation is completed, we can fill in the other figures in the life table (Table 7.2). The k-values for predation $k_5 = 0.47$ (= 66 per cent mortality) and parasitism $k_6 = 0.27$ (= 46 per cent mortality) are quite large.

7.5   The accumulation of life tables

With only a single life table we can have no idea of how the different k-values change with time; but, as we saw for the knapweed gall-fly, we can speculate about the possible interactions between mortality factors.

E

By the time we had eight consecutive life tables for the insect on our five oak trees there were obvious changes in the k-values from year to year. Statistical tests at this stage showed that some of the apparent relationships might have happened by chance, but further work usually confirmed them. We were dealing with a system in which the results were repeatable.

7.6   The causes of population change

Table F in the exercises at the end of the book gives a summary of the winter moth data up to 1968. The basic life table figures can be reconstructed from these figures and can be used in the suggested exercises. To simplify the following discussion and the graphical displays we will use only the figures for the first 13 generations.

Figure 7.3A shows on a logarithmic scale the generation curves of larvae and adult winter moth from 1950 to 1962. Figure 7.3B summarizes the life table data for the same period. Because we have calculated the life tables as if each type of mortality acts in succession, the generation mortality is found from

$$K = k_1 + k_2 + k_3 + k_4 + k_5 + k_6$$

Graphs of these seven terms show at once what is the main cause of population change. The mortalities k_2, k_3 and k_4 are all small and changes in their values are insignificant when compared with those of k_1, k_5 and k_6, and of these last three it is clear that change in k_1 mainly accounts for the changes in total mortality K; both the amount and direction of change in k_1 and K are very similar. From this we conclude that winter disappearance, which we measure as k_1, is the *key factor causing population change*. Morris (1959) was the first to use the term key factor in this sense, but later in the same publication he also had another definition; the key factor was the one which was of most use in *predicting population change*. This second definition was connected to a particular kind of statistical analysis which we consider to be unsatisfactory because it does not necessarily reveal the cause of change; so we prefer to adopt Morris's first definition which implies direct causation. In practice there may be periods when first one mortality and then another is the key factor, and we shall see examples of this in Chapters 8 and 9.

The graphs in Fig. 7.3B give us four useful pieces of information. (1) Winter disappearance, represented by k_1, is the key factor mainly

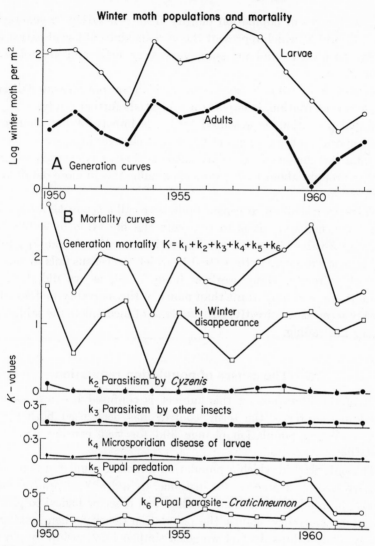

Winter moth populations and mortality

Fig. 7.3

A Winter moth population changes expressed as generation curves for larvae and for adults.

B Changes in the mortality, expressed as k-values, show that the biggest contribution to change in the generation mortality K comes from changes in k_1, the winter disappearance.

causing changes in population density from generation to generation.
(2) k_2, k_3 and k_4 which represent the various kinds of larval parasitism,
so easy to observe and measure, vary very little and are relatively
unimportant.

(3) Parasitism by *Cratichneumon culex*, which was not measured directly
by our census routine, is important and needs further study.

(4) Pupal predation, represented by k_5, is important but changes in
ways opposite to the changes in k_1, so that changes in K are noticeably
less than the changes in k_1. This suggests that k_5 is compensating for
changes brought about by k_1; we will examine this in more detail in the
next section.

This way of looking at census figures we call a *key factor analysis*. It
shows how necessary it is to complete the life table by subtracting
measured *k*-values from the total to estimate *residuals* like k_1 and k_5.
Their size is a measure of the extent to which the census fails to measure
mortality directly. Many workers have found, as we did, that the
residuals are more important than many of the mortality factors which
are easy to measure directly. Any analysis of incomplete life tables may
be very misleading.

7.7   The causes of population regulation

Much of the literature on this subject is confused because, as mentioned in Chapter 5, the term 'control' has been used both for the
process causing population change and for that causing population
regulation.

The only way in which a population might be regulated is by some
negative feed-back process; for instance, by a *density dependent factor*.
Bearing in mind the definition (Section 2.4) it seems logical to plot the
percentage mortality against the population density when looking for
density relationships. In fact we get a simpler relationship by plotting
the *k*-values against the logarithm of the population densities on which
they act; this is a change of scales which in no way affects the definition.
All the *k*-values are plotted in this way in Fig. 7.4. Winter disappearance,
represented by k_1, shows a big variation which is clearly not related to
population density. Pupal predation, k_5, has an increasingly adverse
effect as population density rises; it is apparently a density dependent
factor but the points are scattered about the calculated regression line.

If we wish to know what confidence we can place in the regression slope of $b = 0.35$ calculated in the routine statistical way we run into difficulties. One of these difficulties is that log N is used in the calcula-

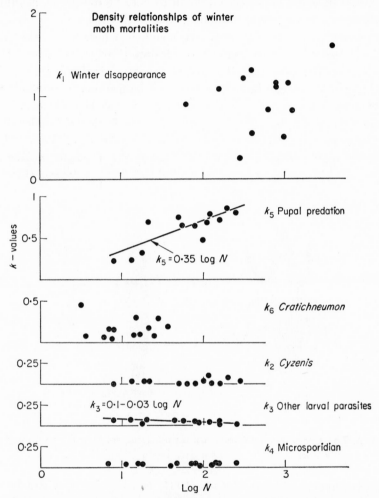

Fig. 7.4 The k-values for the different winter moth mortalities plotted against the population densities on which they acted. k_1 and k_6 are density independent and vary quite a lot; k_2 and k_4 are density independent but are relatively constant; k_3 is weakly inversely density dependent, and k_5 is quite strongly density dependent.

tion of the k-value ($k = \log N - \log S$) so the two measurements are not independent and normal regression and correlation methods should not be used. A second difficulty is that simple regression and correlation statistics require that log N is measured without error; whereas we know that we have sampling errors in our measurements. Those who are interested in seeing how these difficulties may be overcome, should do exercise 8.2.5 at the end of the book. (See Varley & Gradwell 1963, and an example given by Luck 1971.) It is sufficient here to say that valid statistical tests have been applied to this data (Varley & Gradwell 1968) and that we are confident that pupal predation is density dependent and that its effect can be represented by the formula $k_5 = 0.35 \log N$.

When we tried to find what was causing the heavy mortality of winter moth pupae in the soil we put out pupae in the field under sheets of glass, suitably covered so that they were in the dark. We found, by frequent examination, that the carabid beetles *Feronia madida* and *Abax parallelopipedus* and the staphylinid *Philonthus decorus* (Fig. 7.5)

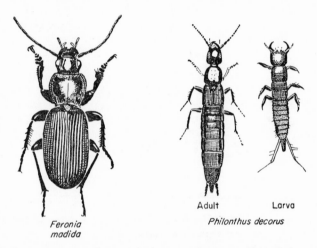

Adult   Larva
Philonthus decorus

*Feronia
madida*

Fig. 7.5 Common predators of winter moth pupae. *Feronia* is a ground beetle (Carabid) and *Philonthus* a rove beetle (Staphylinid).

destroyed many of them. Frank (1967) concluded that these beetles caused more than half of the pupal predation. We also suspected that moles, mice and shrews probably fed on winter moth pupae, and Buckner (1969), who studied the behaviour of shrews in Wytham Wood, concluded that they were the cause of more than half the disappearance of winter moth pupae. Clearly both beetles and mammals are important, but we cannot yet be sure which kills most winter moth pupae. However, on theoretical grounds the k-value caused by a specific predator might be expected to appear as a delayed density dependent mortality: in which case the k-values plotted against $\log N$ would be expected to show a circular or spiral graph when the k-values are joined in a time series. In Fig. 7.6 the successive points for k_5 are joined in sequence and do indeed show some signs of cycling. Shrews, which need to feed every day of the year, cannot depend on winter moth pupae alone, but they may concentrate their efforts in seeking them when the pupae are common. Such a behavioural response from a fairly constant population

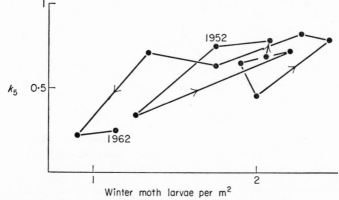

Fig. 7.6 Values for k_5 plotted against the densities of winter moth larvae, with the points joined in a time series. The spiral form of the graph suggests that there is a delayed density dependent component to this mortality.

of shrews might give a strong density dependent effect. The cycling of k_5 is, however, more likely to be the result of changes in the numbers of beetles. The parasite traps in fact caught quite a lot of *Philonthus* adults, but figures graphed in Fig. 7.7 show little sign of a delayed change in relation to winter moth numbers.

7.8 Analysis of parasitism

The question whether parasites act as density dependent factors on winter moth is easily answered from Fig. 7.4. For neither *Cyzenis* nor *Cratichneumon* is there any sign of a density dependent effect on the host. If these parasites had been the only cause of winter moth mortality the picture would probably have been different. Then they would have been key factors and would have been seen to act in a delayed density dependent way. With winter disappearance the key factor, their relatively smaller effect is spread out over a wide range of host

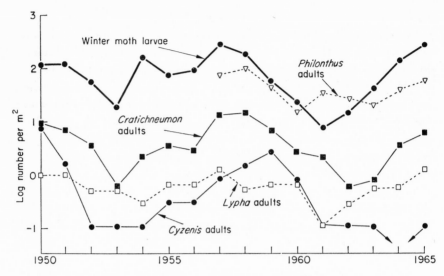

Fig. 7.7 Observed densities of winter moth larvae and adults of its specific parasites *Cyzenis* and *Cratichneumon*, of the non-specific parasite *Lypha* and of the predatory beetle *Philonthus*.

population densities and they respond to the changes in their host's densities rather than cause them.

Surprisingly, miscellaneous parasitism (k_3) is an inverse density dependent mortality with a weak slope of -0.03. The parasites contributing to this mortality are not specific to winter moth and many of them probably have an alternate generation on some other host. One common parasite included here is the tachinid fly *Lypha dubia* whose parasitism of winter moth larvae on oak is clearly inversely density dependent. This parasite regularly attacks three other species of caterpillar on oak as well as winter moth, but in the case of winter moth the parasite larvae are killed by the host's haemocytic defence mechanisms (Cheng 1970). The densities of *Lypha* adults emerging into the parasite traps are plotted in Fig. 7.7 and show that the adult densities of *Lypha* are much more stable than those of *Cyzenis* or *Cratichneumon*. The reason for this relatively greater stability is not known. However, *Lypha* produces a much smaller number of larger eggs than does *Cyzenis* and the relative stability of adult numbers and the inverse density dependent nature of the parasitism by this species may be connected with the low limit to the number of eggs which can be produced; but it is not fully understood.

We were very surprised by the small effect which *Cyzenis* had on winter moth (k_2). After our study began this specific and synchronized parasite was successfully introduced into Canada for the biological control of winter moth there and was remarkably effective (see Chapter 9). Its small effect at Wytham is explained by the very heavy mortality which it suffers whilst in its puparium in the ground. Table 7.3 shows that in 1956 *Cyzenis* pupae suffered a mortality of $k = 1\cdot31$; this represents a 98 per cent mortality suffered by *Cyzenis* between the time of entering the soil as a larva within the winter moth caterpillar and the time of emergence into the traps as an adult. This is much higher than that suffered by winter moth pupae due to predation (k_5 Table 7.2) probably because *Cyzenis* puparia remain in the ground for some four to five months longer than do the winter moth pupae, and so are exposed to predation for a longer time (Hassell 1969a). As well as suffering this pupal mortality we know that *Cyzenis* puparia are themselves attacked by an ichneumonid parasite *Phygadeuon dumetorum* but we have been unable to assess how important this parasite is to the population dynamics of *Cyzenis*.

For *Cratichneumon* we have only a measure of adult density in each year which allows us to assess the minimum percentage mortality of winter moth pupae which they cause. However, from these figures we can estimate the area of discovery for this parasite from formula (4.3). Similarly, we can use the densities of *Cyzenis* adults and the parasitism of winter moth larvae they cause to estimate the area of discovery of this parasite. For both species the areas of discovery a vary considerably from year to year and so do not support Nicholson's theoretical assumption that a is a constant. The quest theory of parasite action (Chapter 4) suggests that the area of discovery should decrease as the density of searching parasite adults increases, and regressions of log area of discovery against log adult parasite density for both *Cyzenis* and *Cratichneumon* show apparently significant negative linear slopes similar to those shown for other parasites in Fig. 4.8. However, once again we are in statistical difficulties. Area of discovery is calculated from the formula

$$a = \frac{1}{P} \log_e \frac{N}{S} \qquad \text{(formula 4.3)}$$

and since P is used in the calculation of a the two measurements are not independent. A method for making a valid statistical test of this relationship is given in exercise 8.2.6, and we will merely say here that the apparent mutual interference of these parasites remains unproven statistically. Nevertheless, it remains true that the inclusion of a mutual interference constant provides a better description of parasite behaviour than the assumption that the parasite's area of discovery is constant at the mean of the observed values.

7.9 A population model for winter moth

The last column of Table 7.4 summarizes the likely properties of the different winter moth mortalities if each acted in isolation. When mortalities with such contrasting and opposing properties act in sequence the outcome must depend on their relative strengths, and we shall study their precise role in the population dynamics of the winter moth by using mathematical population models to represent their actions. How

Table 7.4 Population components used in key-factor analysis of winter moth life tables. Winter moth adults N_A; eggs N_E; full grown larvae N_{L_1}; larvae surviving *Cyzenis* N_{L_2}; surviving other insect parasites N_{L_3}; surviving attack by Microsporidian N_{P_1}; *Cratichneumon culex* adults C.

Name	Formula	Density relationship	Qualitative effect on population
Winter disappearance	$k_1 = \log N_E - \log N_{L_1}$	Big independent variable	Key factor causing change
Parasitism by *Cyzenis*	$k_2 = \log N_{L_1} - \log N_{L_2}$	Weak delayed density dependent factor	Cycles?
Parasitism by other insects	$k_3 = \log N_{L_2} - \log N_{L_3}$ $\simeq 0 \cdot 1 - 0 \cdot 031 \log N_{L_2}$	Weak inverse density dependent factor	Instability
Parasitism by Microsporidian	$k_4 = \log N_{L_3} - \log N_{P_1}$	Effect small	Negligible
Pupal predation	$k_5 = \log N_{P_1} - \log (C + N_A)$ $\simeq 0 \cdot 35 \log N_{L_1}$	Density dependent factor	Stability = regulation
Parasitism by *Cratichneumon*	$k_6 = \log (C + N_L) - \log N_A$	Delayed density dependent factor	Cycles

simple can a model be without losing realism? Fig. 7.3B shows that the model must at least include representations of k_1, k_5, k_6. The main difficulty is that we lack a mathematical description of k_1. Since this key factor cannot be predicted, if we want the model to mimic the observed changes in the population during the period of study, we have to use the observed values of k_1 in the sequence in which they occurred.

We could use the observed inverse density dependent relationship shown in Fig. 7.4 to calculate the mortality caused by miscellaneous parasites (k_3), but for simplicity we add this mortality to that caused by the microsporidian (k_4) and use the mean of the observed values as a constant. These mortalities are so small that omitting them entirely would have little effect on the model.

Mortality due to k_5 is modelled by the density dependent, linear relationship shown in Fig. 7.4. This is a simplification of the real relationship. If we had more knowledge of the predators responsible, and had life tables for them, we would hope to be able to calculate k_5 from measurements of predator density. In the case of the beetles, we would hope to be able to model the way in which their densities would change. As it is, we are unable to mimic the delayed component of this mortality shown in Fig. 7.6, but use only the linear relationship.

The calculated mortalities caused by the specific parasites, k_2 and k_6, are obtained from the relationships between adult parasite densities (P) and the estimated values for their area of discovery a, which is then substituted in the formula

$$k = \frac{aP}{2 \cdot 3} \qquad \text{(formula 4.4)}$$

For *Cratichneumon* the difference between the numbers of winter moth before and after this mortality gives the number of adult parasites in the next generation. But in the case of *Cyzenis*, we have also to calculate the effect of the density dependent mortality suffered by its pupae before we can calculate the number of adults which will emerge to attack the next winter moth generation. The general form of the model for winter moth and its parasites is shown in Table 7.5.

The results of such a model are shown in Fig. 7.8 together with the observed densities of winter moth and its parasites. This model started with the observed densities of winter moth larvae and of *Cyzenis* and *Cratichneumon* adults in 1950 and the different mortalities acting on this generation were calculated using the appropriate sub-models in their

Table 7.5 A model of the interaction between winter moth and its parasites.

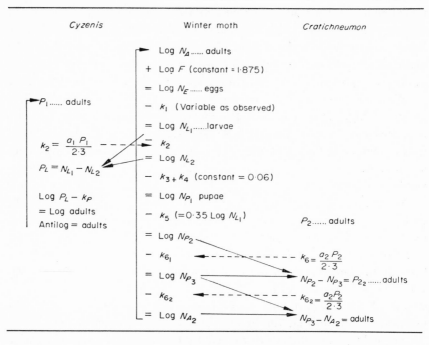

right order. At the end of this generation we calculated the number of eggs laid by the adult moths and this was then reduced by the k_1 value observed in 1951. This cycle of calculations was repeated for each generation. The adult parasite densities used subsequent to 1950 were those calculated to have survived from the number of winter moth parasitized in the previous generation. Thus any errors in the sub-models will accumulate as we calculate more and more generations.

We think that the model shown in Fig. 7.8 fits the observations quite well, and since the sub-models are based on relationships which have a biological as well as a mathematical meaning, we feel that we have gained quite a good understanding of the biology, ecology and population dynamics of the insects concerned.

This model is based on the means of measurements made on five trees. But we know that both *Cyzenis* (Fig. 4.11) and *Cratichneumon* show behavioural responses to the different densities of winter moth larvae on these trees, and also to those which occur on other species of tree in the immediate vicinity. Although we can describe the behavioural responses of the parasites fairly simply, to introduce them in a popula-

tion model involves complications. The host population must be sub-
divided in some suitable way to represent the local concentrations on
different kinds of tree, and from the behavioural response of the parasites, their effects in each subdivision is separately calculated, before
being summed to get the outcome in each generation. Varley & Gradwell
(1971a) reported on a model of this type and we have studied many more

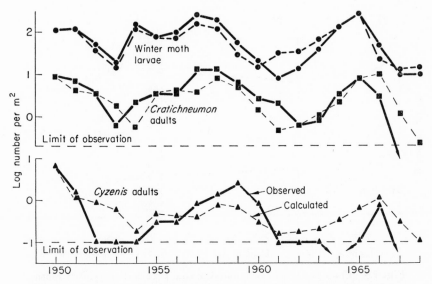

Fig. 7.8 Observed changes in the densities of winter moth and its
parasites (solid lines) plotted beside the densities of these species as
calculated by a mathematical model (broken lines). The method of
calculation of the model is shown in Table 7.5.

variants since then. This has convinced us that the parasite responses to
spatial changes in host density can account for much of the remaining
discrepancies between the observed populations and those calculated
from the very simple models used to prepare Fig. 7.8.

7.10 Predictions using the model

The development of population models is not just an intellectual exercise. When formulated for a pest species the model can provide the basis
for management decisions for biological or integrated control measures.
By suitably altering the parameters in the sub-models we can check
mathematically the consequences of applying additional mortalities

to the pest species or to its parasites or predators and avoid those measures which have no effect on the pest, or an effect the opposite to that intended. This can easily happen; the different mortality factors interact with one another in ways which are too complicated to be predicted except by a mathematical model, as examples using the winter moth model will show.

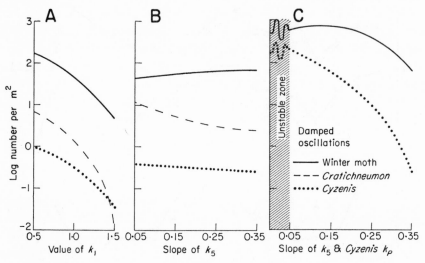

Fig. 7.9 Changes in the calculated mean densities of winter moth and its parasites as parameters of the model used to calculate Fig. 7.8 are changed.

Figure 7.9 shows how the average densities of winter moth larvae and of adult *Cyzenis* and *Cratichneumon* change as different parameters of the model are altered. If each year we used a chemical insecticide to kill either adults or eggs of winter moth, this would increase the average effect of k_1 and reduce the average density of winter moth larvae (Fig. 7.9A). Sticky bands to trap females and winter washes of insecticides to kill eggs are used to reduce winter moth larval populations on fruit trees. Similarly, any measure which decreased the average value of k_1 would result in increased larval densities. We know that k_1 differs between trees. Its high average value is probably because the wide scatter of bud opening times prevents the time of hatch of winter moth eggs from being selected for very precisely. If removal of some trees resulted in those remaining having very similar flushing times, the average k_1 would almost certainly be reduced and the average winter moth population on these trees would be increased. The co-incidence of flushing times would also produce very similar winter moth larval densities on the trees. This, in turn, would reduce the stabilizing effect of the parasites' behavioural responses to different host densities and in consequence the winter moth population densities would be less stable than they are at present.

If chemical sprays were used regularly to kill feeding caterpillars this would have an effect similar to increasing k_1 and would also reduce the average larval density; and if *Cyzenis* and other larval parasites were also killed it would not matter because their present effect is so small. But if wash-off of these insecticides on to the soil killed *Cratichneumon* or the predatory beetles, it would be quite a different matter. The removal of *Cratichneumon* alone from the model doubles the present average density of winter moth larvae, but the effect of removing the predators is more complex. Fig. 7.9B shows that a reduction in the effect of predators by reducing the slope of k_5, would slightly reduce the average winter moth density; this is not the sort of result one would expect from the removal of a potential regulating factor. It occurs because the effect of the predators is more than compensated for the increased density and effectiveness of *Cratichneumon* and, to a much smaller extent, of *Cyzenis*.

If, however, the predators are really responsible for much of the pupal mortality of *Cyzenis*, removal of predators combined with the removal of *Cratichneumon* can lead to a great increase in the average winter moth density. Figure 7.9C shows that as the effect of predators on winter moth and *Cyzenis* is reduced in the absence of *Cratichneumon*, winter moth and *Cyzenis* average densities increase until, with very weak slopes for k_5 and k_P, there is a zone in which the populations of host and parasite show increasing oscillations as in Nicholson's theory.

Before using this model in planning pest management we should like to be sure that the model's predictions are correct. But we studied these populations in a nature reserve where manipulations of populations by the wide use of insecticides cannot be done. However, in Chapter 9 we show how the winter moth model can be altered to fit Canadian conditions. This goes some way to explaining the population interaction in Canada. The fact that it does this can be taken in some small measure to support the validity of the model.

This model for the winter moth has been based on some 13 years data. You may feel that this is too long a period of time to spend on a study in the hope that it may result in a sufficiently good understanding of a problem to be able to base management decisions on it. But no successful short-cuts to an understanding have yet been found and, as you will see in the next chapter, many studies of important pests have continued for much longer periods than this without producing either any real understanding of the problem or any practical solutions to it.

In many of these projects it is technically difficult to get the life table figures, but in other cases once the value of life tables is realized it will be easy to make plans to get the missing census figures. When the winter moth study began in 1949 many of the theoretical ideas had been proposed against which the data were eventually tested. Now that methods for the analysis of census data are becoming more firmly based, we can hope that more field studies will be planned with a view to the construction and analysis of life tables. Only if this is done may ecology develop from a largely descriptive science into one which can take its place alongside the other exact, mathematical sciences.

CHAPTER 8

POPULATION CHANGES OF SOME
FOREST INSECTS

8.1 Synopsis

The numbers of some forest insects change dramatically from generation to generation. Ten-fold increase for three or even five generations may be followed by a similar decline in numbers. Some species have predictable population peaks every nine or 10 years and these cycles seem to be driven either by specific parasites or perhaps by some delayed compensatory change in tree physiology. Other insect species have outbreaks only in stands of mature or of young trees, which are killed.

Bark beetles can overcome tree resistance only by aggregating. They can breed and enormously increase on a tree they have weakened or killed. This produces explosive pest outbreaks whose severity may be increased by the transmission of pathogens like the dutch elm disease.

Both natural and cultivated forests are subject to devastation from outbreaks of specific insect pests. The attack on a dominant tree species helps to maintain the diversity of natural forests.

8.2 Introduction

Forest entomologists are leaders in attempting to understand and predict pest outbreaks. The natural and planted forests they study often occupy extensive areas and persist without drastic change for much longer than do agricultural crops, and for forest insects they have accumulated more detailed records of long-term population changes than exist for any other animal populations. Pest outbreaks are often left to take their course because the use of insecticides may either be impracticable or uneconomic at the current price of timber. The regularity of some of the outbreaks is of great theoretical interest but

135

the interpretation of these long-term records is still very much a matter for speculation. Their correct understanding could help to establish permanent control of the pests, which would have considerable economic consequences. We shall describe here only some of the better documented

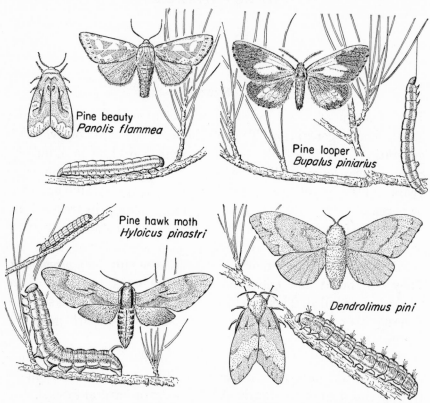

Pine beauty
Panolis flammea

Pine looper
Bupalus piniarius

Pine hawk moth
Hyloicus pinastri

Dendrolimus pini

Fig. 8.1 Four moths whose larvae defoliate pines in Europe. The big hairy larvae of *Dendrolimus* are conspicuous. The larvae of the other three are green with white longitudinal stripes and are hard to see amongst the pine needles.

examples to show the very different patterns of population change which they reveal.

The first detailed information on any species came from the German Forest Service. Extensive planting of pines began as far back as the beginning of the 19th century and insect pests soon became important enough for records to be kept. Schwerdtfeger (1935, 1941) published the back records of the population changes for four kinds of moth whose larvae feed on pine needles. The moths and their larvae are illustrated in Fig. 8.1. They are *Panolis flammea*, a Noctuid, known to British collectors as the pine beauty; *Hyloicus pinastri*, the pine hawk moth; *Bupalus piniarus*, a Geometrid moth known to collectors as the bordered white and to foresters as the pine looper and *Dendrolimus pini*, a Lasiocampid moth, not native to Britain, which is a relative of our oak eggar moth and has a similar large hairy caterpillar. All four species have one generation in the year.

Schwerdtfeger plotted the numbers of larvae per m² of forest floor as a graph against time. Each species occasionally increased to such numbers that the trees were defoliated and food shortage set some limit to their increase; but some small outbreaks terminated without reaching defoliation level. The observed population changes were so large that we prefer to plot them on a logarithmic scale. To prevent the lines from crossing we have plotted each species on its own scale in Fig. 8.2. All

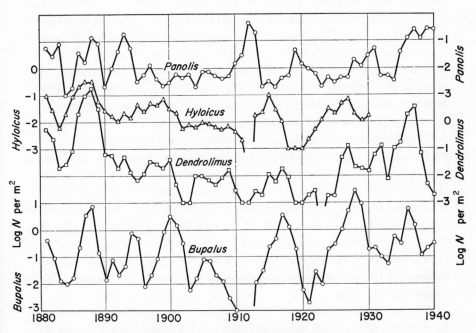

Fig. 8.2 The results of 60 successive census counts of moth pupae or, in the case of *Dendrolimus*, of hibernating larvae per m² of pine forest floor at Letzlingen, Germany. Data from Schwerdtfeger (1935, 1941). After Varley (1949).

four species were exceptionally abundant in 1888, but *Panolis* had an
outbreak in 1912 when the other three species were very scarce. In 1936
and 1937 there was another outbreak in which three of the species were
very abundant, but figures for the pine hawk, *Hyloicus* were not
available.

Schwerdtfeger had no detailed life tables for any of the species but
weather records had been kept and there was also a little information
about parasites. He considered a number of possible explanations for
the changes in numbers. First he rejected the idea that specific parasites
were solely responsible for the ending of the outbreaks because many of
the commonest parasites were not specific and were known to have
alternate hosts on other food plants. Varley (1949) reviewed this study
and compared Schwerdtfeger's figures with what would have been
expected had Nicholson's theory correctly described parasite behaviour.
The observed form of the generation curves in Fig. 8.2 differs from that
we have seen in theoretical curves where parasitism is the key factor,
such as Figs. 4.4, 4.6, 4.9, 4.10. The curve for *Bupalus* in Fig. 8.2
resembles the regular theoretical cycles more closely than do the curves
for the other three, but the peaks are unequal. Furthermore, the peaks
and minima do not seem symmetrical about a fixed mean as are the
theoretical curves. Secondly, Schwerdtfeger considered that the population changes were far too great to be accounted for by weather changes;
however, where population changes for a number of speces were synchro-
nous, as in 1888 or 1936, perhaps weather might have had some indirect
influence in determining the favourability of some other factor. Thirdly,
he found that overpopulation could not always explain the decline in
numbers after an outbreak; many outbreaks declined whilst the population was still too small to have caused defoliation. Schwerdtfeger
considered that an explanation must be in terms of the significant parts
of the whole environment which he termed the 'Gradozön'*, but he did
not suggest how we might explain the interactions of its parts. Since
then a possible way to do this has been found through the analysis of life
tables and the study of the properties of the population models which
we can then construct.

The population changes in the German pine forests do not seem to
be random. Population densties increase for three, four or even five

---

* The term 'life system' which is used by Clark *et al.* (1967) seems to have
similarities in usage to Schwerdtfeger's Gradozön.

consecutive years and the range of population change varies from two to five orders of magnitude.

8.3 Population cycles

A. *Bupalus piniarius*, the pine looper caterpillar

The figures for *Bupalus* show fairly regular peaks roughly every eight years. This species became the subject of a very detailed study by

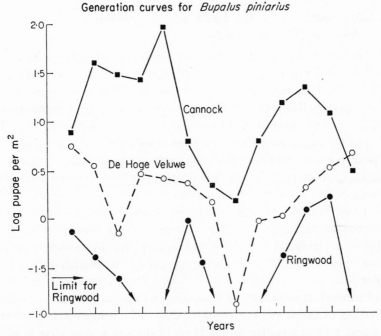

Fig. 8.3 Pupae of the pine looper, *Bupalus piniarius* were counted in samples of litter on the forest floor. The English figures for Cannock and Ringwood represent the highest count in any compartment. The figures for the Veluwe forest in Holland are averages.

Klomp in Holland and there are estimates of numbers from Forestry Commission plantations in England (private communication), which are compared in Fig. 8.3. Although Cannock is in Staffordshire, Ringwood 200 km south in Hampshire and the Veluwe Forest near Arnhem is over 500 km to the east, the range of population change in the three is similar, while the levels about which the changes take place are very different.

Klomp (1966) has published life tables for *Bupalus* for 1950–1965. The key factor causing population change was at first the Tachinid fly *Eucarcelia rutilla* but from 1964 the Ichneumonid wasp, *Poecilostictus cothurnatus*, seems to have been the key factor (pers. comm.). Klomp did not investigate the life tables of these parasites and he has not yet tested the completeness of his explanation with a population model. The changes shown in Fig. 8.3 are not inconsistent with the idea that parasites are an important cause of change.

B. *Malacosoma disstria*, the forest tent caterpillar, in North America (Figs. 8.4 & 8.5)

The forest tent caterpillar is likely to defoliate aspens (*Populus tremuloides*) and other trees in the USA every few years. In any one area, such as the State of Minnesota (Duncan & Hodson 1958) an outbreak begins locally, spreads like a devastating fire over a fairly large proportion of the state and diminishes again to negligible proportions (Fig. 8.4). The timing of the outbreak differs by a year or so from one place to another, but peak populations covered the biggest area in the three years 1951–1953. Long term records of a less precise kind have been summarized by Hodson (1941) for 23 different areas of the American continent. The top half of Fig. 8.5 plots the records for two areas in the east, for Minnesota in the centre and for western Canada from 1885 to 1940. In different places the peaks are not in step, but the average interval between peaks is eight to 12 years depending on whether some of the smaller peaks are counted or not. Although these reports do not include actual numerical estimates, the populations reached the obvious limit, defoliation of the trees. We consider that it is firmly established that the forest tent caterpillar populations change in a way that is cyclic rather than random. Witter, Kulman & Hodson (1972) have published life tables for two successive generations during an outbreak. They found that 100-fold increase was possible, so that a k-value for the generation mortality would be $K = 2 \cdot 0$ if the population was stable. They found a figure of $K = 2 \cdot 245$, and the population declined to about half. The biggest k-values were $0 \cdot 58$ ($= 74$ per cent mortality) for pupal parasitism by the parasitic fly *Sarcophaga aldrichi* and $k = 0 \cdot 64$ ($= 77$ per cent mortality) for starvation of the late larvae combined with adult emigration. *Sarcophaga aldrichi* is known as a regular and fairly specific parasite of the forest tent caterpillar, and Hodson (1941) reported more than 99 per cent parasitism in some places in 1938. The adult flies are strong fliers and Hodson often observed them as far as several miles from the nearest area infested by the host. Parasite emigration probably explains the synchronization of the ending of outbreaks over large areas.

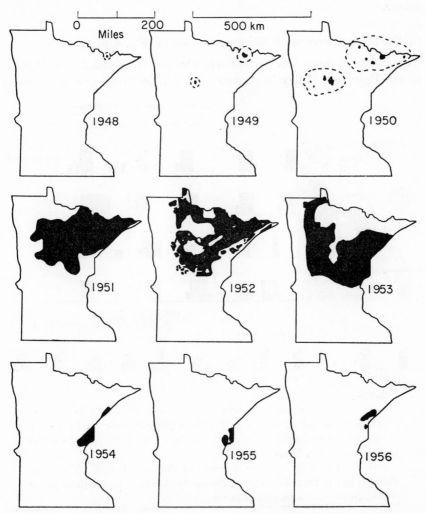

Fig. 8.4 The expansion and subsequent collapse of an outbreak of the forest tent caterpillar, *Malacosoma disstria*, in the State of Minnesota, USA. After Duncan & Hodson (1956).

We know that the interaction between a specific parasite and its host may lead to population cycles of the right period—eight to 10 years—and it is tempting to suppose that these cycles are being driven by parasitism. The crucial test will be to compare the life tables for an increasing and a decreasing population, or if life tables are available for a whole cycle, find if mortality caused by the larva of *Sarcophaga* is the key factor.

C. The larch tortrix moth *Zeiraphera diniana* (Fig. 8.5)

This holarctic moth has particularly spectacular population outbreaks in the Engadine Valley in Switzerland. Within fairly narrow limits of

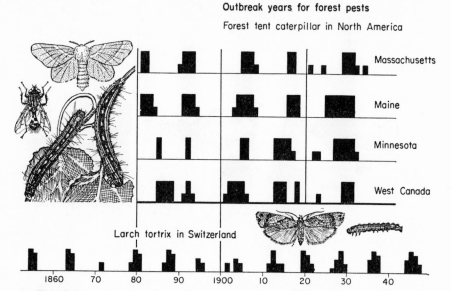

Fig. 8.5
Above Years in which there were outbreaks of the forest tent caterpillar, *Malacosoma disstria* in four parts of North America. Inset: the moth, two caterpillars and the fly *Sarcophaga* a common parasite of the larvae. Data from Hodson (1941).
Below Remarkably regular outbreaks of the larch tortrix, *Zeiraphera diniana*, in the Engadine Valley, Switzerland. Data from Baltensweiler (1964). Inset: the moth and its larva.

altitude on the alpine slopes the larches are defoliated every few years. Baltensweiler (1964) has found records of outbreaks dating from 1854 (Fig. 8.5) which are striking in their regularity. Baltensweiler (1968) and Auer (1968) have published detailed generation curves for the last two population cycles of this species. There are 10-fold rates of change per year during both the increasing and decreasing phases of each cycle.

If a single factor drives the cycles of this moth then we know that its k-value must vary by 2 from one phase of the cycle to another. Auer provided some life table information which included figures for larval parasitism, but the k-values obtained from his figures are never more than 0·6. In seeking an explanation Auer used multiple regression methods but put only part of the life table figures into the regression analysis. When Varley & Gradwell (1970) inserted the missing figures (which were easily estimated by subtraction) they found that it was in this unstudied part of the life table that the key factor lay. Baltensweiler (1968) found that fecundity varied in response to changes in food quality when the trees were defoliated. We shall understand the nature of the forces driving the cycle only after more detailed work on mortality of all stages, including the eggs, young larvae, pupae and adults.

D. *Acleris variana*, the black-headed budworm, Tortricidae (Fig. 8.6)

In the forests of New Brunswick, Morris (1959) studied this tortricid moth, whose larvae feed on both fir and spruce. His analysis of its populations was one of the first which suggested that parasites were a cause of cyclic changes. Of course there is as yet no agreement as to how many cycles must be observed and how regular they must be before the term

Table 8.1 Census figures for the black-headed budworm *Acleris variana* (Lep. Tortricidae) on fir and spruce in New Brunswick. From Morris (1959).

Generation	1	2	3	4	5	6	7	8	9	10	11	12
Number per sq. ft. of foliage	22	112	533	225	12	3·1	3·3	31	150	237	300	183
Percent parasitism	7	9	43	97	88	31	?	14	10	28	44	—

should strictly be applied. Morris published census figures for 12 generations of the larvae from 1946–1958 and measured also the percentage of the larvae which were parasitized (Table 8.1).

Morris analysed the relation between the logarithm of the caterpillar population ($\log N_{n+1}$) and the logarithm of the survival from parasitism in the previous generation ($\log S_n$) which we prefer to express as the

k-value, k_p. He found a high correlation between them ($r = 0\cdot93$). He claimed that the percentage of parasitism could be used to predict the next population and he regarded parasitism as a key factor. He found evidence that it was a delayed density dependent factor, because, when

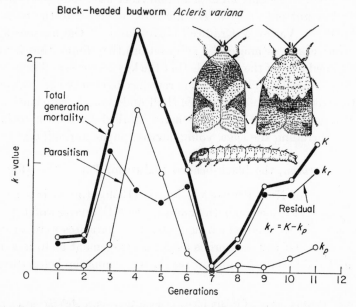

Fig. 8.6 Key factor analysis for the black-headed budworm, *Acleris variana*, in New Brunswick, Canada. Changes in the residual mortality k_r provide a better explanation of the changes in the generation mortality K than do measured changes in parasitism k_p. Data from Morris (1959). Inset: the budworm and two of the many forms of the polymorphic moth.

he plotted log S_n against log N_n, a line which joined the points in sequence formed a figure rather like that in Fig. 4.5

This analysis is not entirely satisfactory because Morris considered only a part of the mortality—parasitism—in his correlation. We need to estimate the residual mortality, k_r which was not measured directly, by substracting the k-value for parasitism from the total K for the generation. The generation mortality from larva to larva, K', the mortality caused by parasitism k_p and the residual k_r are plotted in Fig. 8.6. Although the shape of the curve for k_p has a peak in the right year, the value of the residual k_r is the greater in all but two of the years; the residual mortality provides a better biological explanation of the changes in the generation mortality K than does measured parasitism! Its nature remains unknown.

E. *Choristoneura fumiferana*, the spruce budworm (Fig. 8.7)

Larvae of this Tortricid moth excavate the terminal and lateral buds of various conifers. Its population outbreaks in the Maritime Provinces of eastern Canada are less frequent than those of the larch tortrix moth but even more devastating because the trees are killed. Blais (1965) found records of severe damage to white pines and other conifers in Quebec in 1704, 1748, 1808, 1834, 1910 and from 1947. For this latter outbreak we have a lot of information much of which is summarized by Morris (1963). When the 1947 outbreak had already killed the spruces and balsam firs in parts of New Brunswick it was realized from previous experience that unless something was done vast tracts of forest would be killed.

In New Brunswick the original forest was a mixture of spruce and balsam fir with a scattering of birch. Beside the rivers were much larger white pines, whose valuable timber was exploited first partly because it was easy to float out down the waterways. Then, in this century, the spruces were increasingly taken to make paper. The balsam fir, whose resin interfered with the old pulping process, was rejected and left standing, so that the forest regenerated 'naturally' mainly as a mixture of balsam fir and birch. Then in the early 'forties the birches suffered from 'dieback' and died over vast areas. Whether there is any causal connection between this reduction in diversity and the 1947 outbreak which followed we cannot be sure, but Voûte (1946, 1964) showed that pest outbreaks in forests are usually less severe in mixed woodland of uneven age than in even-aged plantations.

The Canadian Forest Service surveyed the pest problem from the air and from the ground. From the air, areas of defoliation were easily distinguished by their brown colour from green healthy trees, but healthy trees had to be studied from the ground and sampled for eggs of the budworm to determine whether the district required insecticide treatment. A big scientific team did the counting and some hundreds of light aircraft flew from special airstrips throughout the forest to spray DDT on to the areas where trees were endangered. At the height of the spray programme in 1957 some 5·2 million acres were sprayed; this was only 40 per cent of the total area known to be suffering damage. The DDT

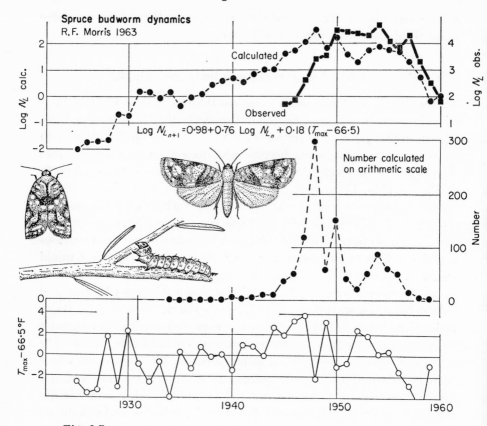

Fig. 8.7
Above The observed changes in the numbers of the spruce budworm, *Choristoneura fumiferana*, and the changes calculated using formula 8.1. Note the different scales for observed and calculated.
Middle The calculated numbers on an arithmetic scale.
Bottom The changes in mean temperature used in the calculations. Data from Morris (1963). Inset: the budworm and adults of *C. fumiferana*.

killed enough of the budworm to keep the trees alive, but seems to have contributed little to bringing the outbreak to an end. It was only when budworm populations in the unsprayed areas were seen to be collapsing from some natural causes that spraying could be ended.

Morris (1963) reported on the population dynamics of the spruce budworm. For many areas he had life tables but, as in the work on its relative in the Engadine, parasites were not studied in enough detail to produce life tables for them. So an interpretation on the lines we used for the winter moth was not possible and regression methods were used instead, on the assumption that important population changes from year to year, 'although *affected* by many factors, may be essentially *determined* by a few "key" factors'. Sub-models were built up step by step for the survival in each age interval and the variance between the outcome of the model and observation was checked at each stage.

Morris did not in fact use these sub-models in his final 'predictive equation' which was based on key-factor analysis over the whole life cycle. He found that the simple equation which best fitted the data was

$$\log N_{n+1} = 0\cdot98 + 0\cdot76 \log N_n + 0\cdot18 \, (T_{\max} - 66\cdot5) \qquad (8.1)$$

In this equation the $0\cdot98$ represents the reproductive increase of the population N which is almost 10-fold. When this is added to $\log N_n$ the sum represents the number of offspring. The next term in effect includes a density dependent factor $k = 0\cdot24 \log N_n$ subtracted from $\log N_n$. This is similar to our representation of $k_5 = 0\cdot35 \log N_n$ for pupal predation of winter moth in Chapter 7. The last term includes the mean daily maximum temperature in degrees Fahrenheit $T_{\max}$; and shows that the logarithmic increase rises with mean temperature, the population doubling or halving depending on whether $T_{\max}$ rises or falls by $1\cdot7°$F.

Morris found that this equation explained 68 per cent of the variance in the population, which he regarded as surprisingly high in view of the other sources of variation in the data. However, when the outcome of this formula is compared with the census figures in Fig. 8.7 the discrepancies seem considerable. The predicted population peak is in 1948, with a second maximum half its size in 1950. Observations showed a peak in 1950 but the largest one in 1954. The observed population increase was some four years later than that predicted by equation (8.1) and the fall about a year later. These discrepancies are larger than those between the winter moth model and the observed values (Fig. 7.8). Both models use one observed figure each year, $T_{\max}$ in equation (8.1) and k_1 as observed in the winter moth model, so the models are comparablein this. We shall consider these methods further in the discussion.

8.4   Introduced Forest Pests

With the deliberate introduction of ornamental and economic forest and fruit trees a number of pests have been introduced into areas where they have found climatic conditions to their liking and where native trees have provided food. We will mention only a few instances here. Another —the winter moth in Nova Scotia—is referred to in the last chapter.

The Gipsy Moth (*Lymantria dispar* L. (Lep. Lymantriidae)) was introduced accidentally into the United States about 1868 and rapidly increased to plague proportions, causing defoliation of the shade trees in many cities of eastern USA. Campbell (1967) provides an explanation of the population changes, but his census for the parasites (many of which have been introduced in the hope of achieving biological control of the pest) is not detailed enough to enable him to model their activities in a realistic way.

The European pine saw-fly, *Neodiprion sertifer* (Hym. Tenthredinidae) is a serious pest in Europe, particularly of young trees in plantations. Since its accidental introduction into North America it has become equally serious in its attack on native pines in the north-east and is still extending its range there. A nuclear polyhedrosis virus, first found in Sweden, has been artificially spread to combat this pest. This can cause high mortality of saw-fly larvae whenever high populations of this species are damaging young trees.

Matsucoccus feytaudi (Hemiptera, Coccidae) is a minute scale insect, discovered as a new species infesting the bark of the maritime pine, *Pinus pinaster*, the dominant forest tree of the Maures in Southern France. Its origin is unknown, but it might have been introduced with the many exotic conifers grown in estates and gardens. Its effect on the maritime pine is devastating. Large areas, once dominated by the pine, are now transformed into oak forest—the oaks being previously sub-dominants. It is not the scale insect alone which produces the effect; the scale weakens the trees which are then susceptible to the effects of bark beetles, which complete the destruction (Schvester 1971).

8.5   Pests which aggregate

Bark beetles are notorious because of the severity and localization of their damage. Many species damage the trees in two quite different ways: first the immature beetles feed on twigs and foliage. The tree has no special resistance to this kind of damage and is weakened physiologically if the damage is extensive. Second the female beetles excavate maternal galleries in the cambium layer of the trunk and larger branches and lay their eggs there. Healthy trees flood these galleries with sap or resin and the attack fails, but if the tree is weakened physiologically by damage of the first kind, or by unfavourable weather conditions, then it is unable to resist being overwhelmed by the breeding beetles. Trees which have been blown down by the wind or trees stacked at saw mills before being sawn up, are readily colonized and the resulting populations of adult beetles may in the next generation feed on normal trees nearby and weaken them until they are also suitable for maternal galleries. Beaver (1966) studied the dynamics of the elm bark beetle *Scolytus scolytus* in billets of timber and to describe the survival used graphs in a new way which partly overcame the difficulty of describing what happens in overlapping generations of the beetle larvae.

The Western pine beetle *Dendroctonus brevicomis* has well developed powers of aggregation. The females produce a pheromone, frontalin, whose scent is attractive to other males and females, so they can concentrate their attack on one tree; then the weakened tree becomes the focus for an explosive local outbreak, which can spread to other trees and kill them (McNew 1971).

For such insects the mean population density has little meaning. It is the concentration on each tree which matters and the effect depends on the tree's physiological responses to other environmental factors.

8.6   The transmission of pathogens by insects to trees

Insects can interact with trees by transmitting parasitic fungi or virus diseases. The dutch elm disease, caused now by a virulent strain of the fungus *Ceratocystis ulmi*, has suddenly increased in parts of Southern England where, by 1972, a large proportion of the native elms in hedge-rows and woodlands had been killed. It is spread from tree to tree by adult bark beetles of the genus *Scolytus*, which breed very successfully in newly dead or dying trees, from which the emerging adults can readily carry the infection to healthy trees nearby. This explains the catastrophic contagious spread of the disease.

Siricid wood wasps also transmit the spores of specific pathogenic fungi which they store in a special organ associated with the ovipositor. The larvae feed on the mycelium. The Alder Woodwasp, *Xiphidria camelus*, transmits the fungus *Daldinia* to alders. Thompson & Skinner (1960) have provided an excellent film of the life history of *Xiphidria* and of three kinds of hymenopterous parasite and hyperparasite. The wood wasp, *Urocerus*, transmits *Stereum sanguinolentum* to spruces and has become a serious pest in Australia. Its population dynamics and the possibility of biological control are currently being studied.

8.7 Discussion

The conclusions from this chapter are often tentative because of inadequate census figures, but clearly we are not studying one single phenomenon for which a single explanation will eventually prevail. The local outbreaks of the bark beetles tend to spread from some focus, which may be some wind blown trees, or a pile of cut logs beside a saw-mill. In the case of the western pine beetle these effects, though present, are overshadowed by the ability of the insect to aggregate by means of pheromones. In all these cases the infestation spreads very rapidly because at an early stage the local aggregation of the insects turns the excess of protected food into a usable resource. In effect these changes in the usable food supply acts as an inverse density dependent factor which is also the key factor. The food supply is soon exhausted and the next generation moves on elsewhere.

For species which show cyclic population changes, like the pine looper, larch tortrix, forest tent caterpillar and black-headed budworm, the factor driving the cycles must be a delayed density dependent factor. Parasites could act in this way, but when you have answered question 8.2.1(a) you will see that what might well be cyclic changes in the winter moth population were caused by some component of k_1, the winter moth disappearance; the evidence was against parasitism playing any significant part in this. There is suggestive evidence that the cycles in the forest tent caterpillar are driven by *Sarcophaga* but this does not yet amount to proof. That parasitism plays a part in the cycling of the larch tortrix in the Engadine and the black-headed bud-worm in New Brunswick is proven but what really drives the cycles is unknown. In all cases the observed parasitism is correlated with the change in the generation mortality, but the parasitism measured is insufficient to be the major cause of change. Either there is some unmeasured pupal parasitism or predation, or the cycle might be driven by changes in tree physiology. White (1969) considers that 'stress' may have important effects on tree physiology. This stress can arise either from insect damage or from unfavourable weather conditions such as summer drought or winter flooding. To provide an explanation for the regular cycle of changes in the insect populations the physiological change in the tree would have to persist for more than a year after the stress which originally caused it. We shall understand these interactions only when stress is studied experimentally and changes in plant physiology, such as the nutrient content of the tree sap, are measured and correlated with changes in insect growth or survival.

When we think back over the examples of field studies in the last few chapters, none offers an explanation of population change or population level which is entirely satisfying, but we have made a beginning. We understand some of the complex interplay between field populations of insects, their parasites, predators and the weather. The progress made in these very different studies has depended very much on the amount of detail provided by the census method chosen. For winter moth and some of its parasites we had two counts in each generation so the life tables could be completed with two residuals—k_1 and k_5. However, we have failed to model the observed changes in k_1 which we expected might be influenced by weather conditions. Our sub-model for predation (k_5) is over-simple, but we had no good census figures for the predators concerned except during a brief period (Frank 1967, Buckner 1969). Nevertheless in spite of the simplifications and omissions involved the outcome of the model fitted observation rather well. For the spruce budworm Morris had partial life tables based on one count in the year from many localites but apparently had no continuous set of complete life tables from one place. Without any measurement of parasite survival, life tables for the parasites could not be constructed. We suspect that the lack of any representation of the effects of parasites may be the reason for the rather poor fit of Morris's model to the census counts, because the precise way in which the effect of parasites is introduced into a model makes a big difference to the outcome. At an early stage in modelling winter moth we contrasted three models, one in which the parasite was given an area of discovery equal to the mean value observed, one model with half and one with twice this value (Varley & Gradwell 1968). The extent of population change in the winter moth was very different in these models and a model without parasitism fitted very poorly indeed. We consider that it is extremely important to find ways to measure the effects of parasites and predators, and to get life table information for them (Varley & Gradwell 1971b).

We have seen that very different kinds of analysis have been used to explain the causes of population change and of population density. Here we do not wish to go far into the statistical jungle which has grown up in this controversial field, but we feel that we must mention some of the basic difficulties.

One of the methods which has often been used is multiple regression, for which the basic formula can be written

$$Y = b_0 + b_1 x_1 + b_2 x_2 + b_3 x_3 + \ldots \tag{8.2}$$

where the x terms are independent variables and the b terms are constants. The method was devised by R.A.Fisher to quantify the additive effects of combinations of fertilizers on the yield Y of a crop. He already knew that the effects were linear and additive. It is now sometimes used for a very different purpose—to discover if either population density or population change is correlated with any measurements which may be available to serve as independent variables.

It is important to realize that this technique is a means of getting a predictive equation. Whether or not the equation is useful will depend upon the accuracy with which it predicts. It is *not* a method that is likely to discover what particular biological mechanisms operate. Its use for this purpose may lead to erroneous conclusions. We have seen how Davidson & Andrewartha (1948a, b) and Auer (1968) both used multiple regression methods and both overlooked the biological importance of major mortalities. Other difficulties in the use of multiple regression to predict population density or change arise if there are any interactions between the independent variables or if the different terms are not additive. Principal component analysis may be used to overcome some of these difficulties, but the problem is too complex for us to define any precise rules here about what methods should be used on field data.

We feel a more biological approach is to use as the basic formula for population change from one generation to the next:

$$\log N_{n+1} = \log N_n + \log F - K_n \tag{8.3}$$

and

$$K_n = k_1 + k_2 + k_3 + k_4 + k_5 + k_6 \tag{8.4}$$

where k_1 etc are k-values referring to generation n, K_n is the generation mortality of generation n, and F the rate of increase. The graphical key factor analysis of Varley & Gradwell makes use of formula (8.4) which is a valid representation because the k-values have been calculated assuming that they add up to K.

If biologically realistic models for the individual k-values can be developed, then the combination of formulae (8.3) and (8.4) will give a population model which will predict both population change and population density; and at the same time provide an explanation for the population change and for the population level. As we said in Chapter 7, such models are the only ones likely to be of use when attempting to make decisions on pest management.

Ideally the research workers should plan methods in relation to the statistical or graphical methods they intend to employ in the analysis of their results. We suggest strongly that they should test these methods and try them first on a deterministic model which matches as well as possible the kind of system which they imagine they are investigating in the field. Luck (1971) used this approach in comparing methods of testing for density dependency and further tests can be made on the figures we give in the exercises for Chapters 7 and 8.

CHAPTER 9

BIOLOGICAL CONTROL

9.1 Synopsis

The classic examples of biological control are the introduction of a predatory beetle and a parasitic fly from Australia into California to combat the ravages of the cottony cushion scale on *Citrus* and the importation from S. America to Australia of a moth to destroy the aggressive cactus, *Opuntia*. In these and many other successful projects an introduced species of pest or a weed had become far commoner than in its native home. The abundance of the pest is permanently reduced by adding one extra link to the food chain.

Many introductions have failed; sometimes this is because the introduced beneficial species could not stand the climate, sometimes because its life history failed to synchronize with that of the pest. Successes have been more frequent with pests of orchard and tree crops than with agricultural crops.

The winter moth was introduced into Eastern Canada and became a pest, but the importation and liberation of the parasitic fly *Cyzenis* from Europe has now reduced its numbers to a low level, so that trees are no longer defoliated. The changes in numbers of both pest and parasites were recorded. By combining with the Canadian observations the new theoretical ideas from Chapter 4 and the information about winter moth and its parasites in England in Chapter 7, we have been able to find a mathematical model which imitates the observed events quite well.

Recent trends in biological control seek to exploit our understanding of the factors limiting the efficiency of beneficial insects and improve their performance by proper management. Management programmes will be easier to devise if the complex interactions which are involved can be adequately described by a mathematical model.

154

9.2   Introduction

In this final chapter we shall consider the practical application of some of the ideas introduced in the previous chapters, that is, the use of natural enemies in pest control. The object of pest control is to reduce the numbers of pests present and hence the damage they cause. Normally this will increase the profit on the crop in question, unless the costs of these control measures are greater than the resulting benefits. Clearly the term 'control' in this applied sense is rather loosely used. It is not synonymous with 'regulation'; an insect population may be no more stable after successful control than previously when of pest status. For example, a pest population may fluctuate between 10 and 1000 individuals per unit area prior to its control and between 1 and 100 per unit area afterwards. With 100-fold changes in each case, the stability of this population has not changed, only the level about which the fluctuations occur.

The potential of biological methods in pest control is immense. All major pests are attacked by a variety of natural enemies and these may be used as a primary means of control or as part of an integrated control programme. It is the aim of *biological control* to manipulate these natural enemies (parasites, predators and pathogens) in an attempt to reduce the pest numbers and keep them at much reduced levels. The 'manipulation' usually involves the introduction of natural enemies into a region where they previously did not exist to counter accidentally introduced pests of introduced crops. A typical 'classical' biological control programme of this kind involves the introduction of relatively few individuals of a species of parasite (or predator), to control a very abundant host. If the introduction is successful, the parasite population will increase rapidly causing increasing percentages of hosts to be parasitized. This in turn leads to a decline in the host population and a subsequent decrease in the parasite population. Ideally, both host and parasite should then co-exist in a relatively stable interaction at very much reduced population densities. Apart from this use of *exotic* natural enemies there is now also very interesting work being done to try to improve the efficiency of *indigenous* natural enemies in controlling pest populations (see below).

9.3   Successful control of pests and weeds

The first and most famous example of biological control was the control of a citrus pest, the cottony cushion scale (*Icerya purchasi*) by a coccinellid

beetle, *Rodolia* (= *Vedalia*) *cardinalis* (Fig. 9.1A). The cottony
cushion scale is native to Australia where its original food plants were
various native acacias and other trees from which it later spread to the

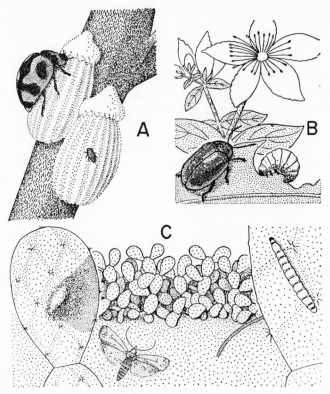

Fig. 9.1 Some successful agents of biological control.
A The ladybird beetle *Rodolia* and its larva which feed on a citrus
pest, the cottony cushion scale insect *Icerya*.
B The chrysomelid beetle *Chrysolina* and its larva both eat the Perforate St John's wort *Hypericum*.
C The moth *Cactoblastis* whose caterpillars feed on the prickly pear
cacti *Opuntia*.

introduced *Citrus*. During the later parts of the 19th century, citrus
plants were imported into California from Australia, and with the citrus
came the cottony cushion scale without its normal complement of
natural enemies. The scale insects thrived in the Californian climate and
by 1887 had become a threat to the whole of the newly-founded citrus
industry. The United States Department of Agriculture then arranged

that Alfred Koebele should visit Australia to seek the natural enemies of *Icerya* in a region where it does not have pest status. He discovered that the scale in Australia and New Zealand is attacked by a variety of beetles, flies and Hymenoptera (Koebele 1890). Two of these, a parasitic fly, *Cryptochaetum iceryae*, and the vedalia beetle (*Rodolia*) were liberated in California in areas where the scale insects were already very numerous. Both species became established, but in most areas the beetles were the more successful of the two. The female beetles lay large numbers of eggs beside and under scales. The larvae then feed avidly on the scale insect eggs, or on other stages up to the adult, and complete their development in as little as 30 days. From over 500 beetles liberated in 1889 on heavily infested trees in California, the spread was so rapid that soon the cottony cushion scale as a pest was but a memory. Both species still exist in Southern California, but at extremely low levels of abundance. Serious outbreaks of cottony cushion scale now occur only where insecticides have been used against other citrus pests such as the citricola scale. The insecticide eliminates the *Rodolia* beetles, and the cottony cushion scale can then increase enough to harm the trees.

Most cases of successful biological control have followed a similar pattern.

1 The search for natural enemies in the country of origin of the pest, especially in climatic zones similar to the planned liberation point.

2 The importation of one or more of these natural enemies. These are then studied under strict quarantine and screened for possible harmful effects on native beneficial insects or crops. There is often large-scale breeding prior to release.

3 The establishment of one or more of the released parasites or predators, their increase in numbers and the subsequent decline in the pest population.

4 The stable co-existence of host and parasite or predator at very much reduced densities.

Such control programmes, where successful, are remarkably economical. The total cost of Koebele's expedition was less than 5000 dollars, yet it achieved the permanent elimination of the cottony cushion scale infestation in South California. Costs are, of course, much higher nowadays, but still very much lower than the four million dollars or so that are required to put a new insecticide on the market, let alone apply it regularly in order to achieve continuous 'control'. De Bach *et al.*

(1971) estimate that in California alone about 200 million dollars have been saved during the past 45 years as a result of successful biological control projects.

Biological control has also been used against noxious weeds, a classical example being the prickly pear cacti (*Opuntia inermis* and *O. stricta*). These were introduced into Australia as desirable hedge-row plants in the early part of this century. They spread rapidly and formed impenetrable thickets on large areas of previously useful land in Queensland and New South Wales. An expedition was sent to South America where several species of moth whose larvae feed on the fleshy cladodes of the prickly pear were collected and shipped to Australia. One of these, *Cactoblastis cactorum* (Fig. 9.1c), rapidly became established and caused a large-scale decline in the prickly pear. Scattered plants of the prickly pear are still found over much of its previous range in Australia but it is no longer dominant and much of the natural vegetation has returned.

These examples show that biological control is concerned with decreasing the numbers of a pest (a primary consumer in the case of phytophagous pest, or a primary producer in the case of a weed) by increasing the numbers and effectiveness of the animals feeding on the pest (secondary consumer or primary consumer). This is illustrated in a very simplified form in Fig. 9.2 (Varley 1959).

In Fig. 9.2A we have the situation where a plant has become dominant because it is free from primary consumers or competitors. Such was the case of the prickly pear in Australia prior to the introduction of *Cactoblastis*. The situation in Fig. 9.2B illustrates the result of the

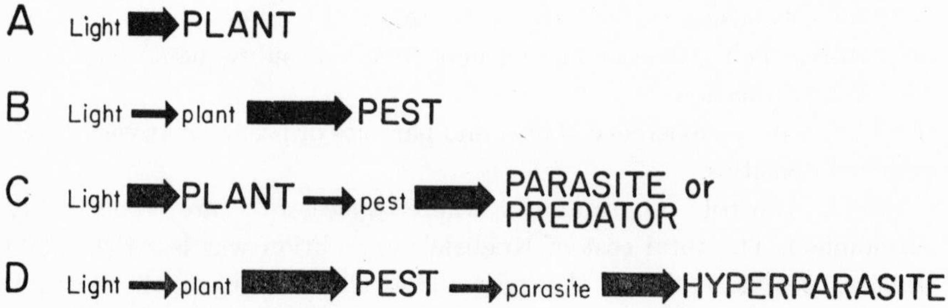

A Light ➡ PLANT

B Light → plant ➡ PEST

C Light ➡ PLANT → pest ➡ PARASITE or PREDATOR

D Light → plant ➡ PEST → parasite ➡ HYPERPARASITE

Fig. 9.2 Simple energy chains in which the thick arrows point to dominant species which have limiting resources and the thin arrows to species which are limited by their consumers and thus have superabundant food which they cannot fully utilize. From Varley (1959).

establishment of a pest species such as the cottony cushion scale (or in the case of the prickly pear, the introduction of *Cactoblastis* from South America). The plant is now no longer dominant and it is the herbivore which is limited by lack of food or available energy. Situation 9.2c represents the successful biological control of an insect pest such as the cottony cushion scale by *Rodolia*. The pest is now limited by the parasite or predator, which once again allows the plant to increase to the limit of its energy supply. Figure 9.2d illustrates the danger of a hyper-parasite becoming established where biological control has been effective. The pest is once again able to increase at the expense of the plant.

9.4 Successes and failures

Although the majority of introduced natural enemies have failed to produce significant control, there have also been many successes. Many of these have been in North America. For example, Turnbull & Chant (1961) review 31 biological projects in Canada, of which they rate 12 as successful and one as partly successful. Clausen (1956) reviewed the effectiveness of biological control projects in the USA up to 1950 (see Table 9.1A). One hundred and ten beneficial species were established

Table 9.1A The effectiveness of biological control projects in the USA up to 1950. Data from Clausen (1956).

	Fruit crops	Field and garden crops	Forest and ornamental trees	*Total*
No. of pest species	33	42	16	91
No. of pests effectively controlled*	12	2	4	18
No. of beneficial species established	55	20	35	110

* These figures are conservative, and do not include partial and local successes.

in an attempt to 'control' 91 species of pest of which 18 species were completely 'controlled'. More recently several other major pests in North America have been controlled biologically, such as the winter moth in Canada (Embree 1971), the Olive scale (Huffaker *et al.* 1962) and Klamath weed (Huffaker & Kennet 1969) in California. This is a common plant in Britain, known as the Perforate St John's Wort, *Hypericum perforatum.* The beetle *Chrysolina* introduced to California is native to Britain where the plant is never a serious pest (Fig. 9.1B).

Table 9.1B summarizes the world situation up to the end of 1969. Many of the early successes of biological control were in Pacific islands.

Table 9.1B Number of pest insect species subjected to biological control attempts by the importation of natural enemies and the number and degree of successes attained (up to and including 1969). After DeBach (1971).

Number of pest species involved	Number and degree of successes				No. results
	Partial	Substantial	Complete	*Total*	
223	30	48	42	120	103

For example, a chafer beetle from Japan, *Anomala orientalis*, has been controlled by a large wasp, *Scolia manilae*, introduced into Hawaii from the Philippines by Muir in 1916. Muir also introduced into Hawaii a tachinid fly, *Ceromasia*, which has controlled the sugar-cane beetle borer, *Rhabdocnemis* (Fig. 9.3A). Another tachinid, *Ptychomyia remota*, has controlled the coconut moth, *Levuana iridescens*, in Fiji. Also in Fiji, the coconut leafmining beetle, *Promecotheca reichei* (Fig. 9.3B), has been successfully controlled by a Eulophid wasp, *Pleurotropis parvulus*, that was imported from Java (Taylor 1937). Because of such examples, it was formerly widely assumed that biological control is best suited to island conditions. However, the experience with cottony cushion scale and the more recent successes in North America indicate that biological control can be just as effective on continents as on islands.

Where biological control has failed, it is because the beneficial insects introduced have not become established or, if established, have proved to be ineffective. The reasons for such failures are almost always unknown but there is now a body of experience which should make it easier to select more appropriate natural enemies for introduction than has often been the case in the past.

Presumably, unsuitable climatic conditions, differences in host biology compared with the native region and unsuitable cultural practices are all important considerations where parasites have failed to become established. An inadequate taxonomic knowledge of the host or parasite populations may also contribute. This has been particularly true in the case of the California red scale (Fig. 9.3c). Many of the parasites that failed to become established were collected in the Orient

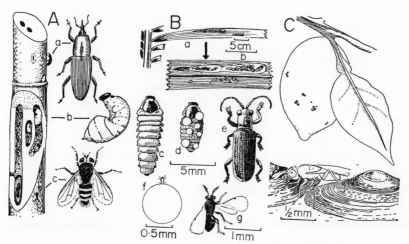

Fig. 9.3
A The tachinid fly *Ceromasia* (c) which has successfully controlled the weevil *Rhabdocnemis* (a) whose larvae (b) bore within sugar cane.
B The Eulophid wasp *Pleurotropis* (g) whose introduction into Fiji successfully controlled the beetle *Promotheca* (e), whose larvae (c) bore in the leaves of the coconut (a, b). An accidental introduction of the mite *Pyemotes* (f) which also feeds on the larvae (d) had previously destroyed the natural control exerted by several other parasitic wasps. (See Section 9.4.3).
C The California red scale on a lemon. Below, magnified, the parasitic wasp *Aphytis* attacks the scale insects.

from scales misidentified as California red scale. Misidentification of parasites has also hindered this project. De Bach *et al.* (1971) report that 'Early in the century, a species of *Aphytis*—in all probability *A. lingnanensis*—was recognized as a common [chalcid] parasite of California red scale in southern China, but no effort was made to introduce it, due to its misidentification as a species already present in California'. Later, after the California species was found to be *A. chrysomphali*, *A. lingnanensis*, which had been confused with it, *was* imported and

proved to be a very effective natural enemy of the scale, displacing
A. chrysomphali from most of its former range.

A high proportion of introduced natural enemies that do become
established fail to reduce their host populations completely below the
level of pest status. Some are partially successful, others completely ineffective. The examples we give below show some of the reasons for this.

9.4.1 *Climate*

We mentioned above that *Aphytis lingnanensis* has been very effective
against the California red scale. However, this has been the case only in
the coastal parts of California. Inland where the climatic conditions are
more extreme the parasite has proved to be much less effective. DeBach
(1965) has shown that the very high temperatures and low humidities
inland in summer and the low winter temperatures in these areas lead
to higher larval mortality, reduced fecundity and unbalanced sex-ratio
compared with the coastal regions. The parasites are not eliminated
from these inland areas because some individuals are genetically more
resistant to the climatic conditions and because survival is higher.
Economic control in the inland areas has now been obtained using *A.
melinus* introduced from India and Pakistan.

9.4.2 *Geographic races*

The whole success of a biological control programme may hinge on the
early recognition of any biological races amongst the introduced parasites, or any differences between the pest populations and populations
of the same species in the parasite's country of origin.

The biological control of olive scale in California shows very well the
importance of distinguishing between different biological races of
parasites. The olive scale became established in California during the
1930's and in the following 30 years has spread throughout the central
olive growing areas of the State. A biological control programme was
initiated in 1949 with the introduction of *Aphytis maculicornis* from
Egypt. During the next few years several shipments of this species were
made from Middle Eastern countries. Amongst these were recognized
four 'biological strains' which although indistinguishable morpholo-
gically had distinct biological characteristics (Doutt 1954). Only one of
these, the 'Persian strain', gave some degree of control in field experiments. The subsequent colonization and release of this strain of *Aphytis*

has been thoroughly described by Huffaker and his co-workers (Huf-faker *et al.* 1962, Huffaker & Kennett 1966), who found that the seasonal climatic conditions greatly affected the efficiency of the parasites. The spring generation of olive scale suffers up to 95 per cent parasitism, but in the very hot and dry summer conditions the parasites are ineffective against the summer generation of the scale.

The situation has now been considerably improved by the establish-ment of a second Chalcid parasite, *Coccophagoides utilis*, which causes between 20 and 60 per cent parasitism in each of the two host genera-tions. It thus serves to prevent the large increases in olive scale during the summer months.

9.4.3 *Host availability*

Parasites with shorter life cycles than their hosts have the advantage of being able to increase relatively more rapidly in numbers than synchronized parasites. If the host's generations are not completely overlapping, however, there is always the danger that such unsynchro-nized parasites will become adult and need hosts when few hosts of a suitable stage are available.

The importance of this to biological control was first shown in the case of the coconut leafmining beetle, *Promecotheca* (Taylor 1937). Under natural conditions the beetle has many overlapping generations and the larvae are heavily parasitized by several chalcid wasps which have much shorter life cycles than that of the beetle. Under these conditions there was little damage to the palms. The situation was dramatically changed when a predatory mite (*Pyemotes = Pediculoides*) (Fig. 9.3B) was accidentally introduced. The mites fed on the beetle larvae and pupae, but not on the adults and eggs, which had the effect of changing the pest from one with overlapping generations to one with each stage more-or-less synchronized. The mites and the native parasites then declined in numbers because they were not synchronized with their host. The wet season is unfavourable to the mites whose numbers then decreased further; so the beetles, now free from mites and parasites, increased rapidly to the limit of their food supply and defoliated the trees. Biological control was finally achieved by the introduction of the Chalcid parasite *Pleurotropis* (Fig. 9.3B) from its native home of Java where it is a parasite of a closely related beetle, which has a life cycle approximately the same length as that of *Promecotheca*.

9.4.4   *Host resistance*

There has been much attention focused recently on the frequent development of resistance by pests to various insecticide compounds.

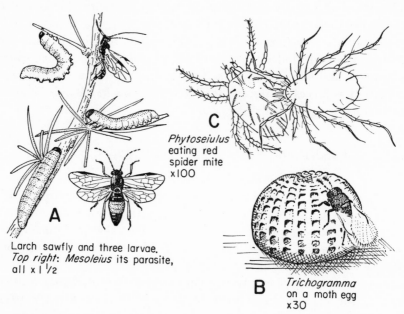

C *Phytoseiulus* eating red spider mite ×100

A Larch sawfly and three larvae. *Top right*: *Mesoleius* its parasite, all × 1 ½

B *Trichogramma* on a moth egg ×30

Fig. 9.4

A The ichneumonid parasite *Mesoleius* was introduced from England to control the saw-fly *Pristiphora* whose larvae feed on larch. After some initial success, the saw-fly larvae became resistant to this parasite.

B Attempts to use species of *Trichogramma*, which parasitize moth eggs, for biological control of Lepidoptera have been largely unsuccessful.

C The predatory mite *Phytoseiulus* has been used successfully to control the red spider mite which attacks cucumbers in glasshouses (see Section 9.6).

The development of host resistance towards a parasite species, however, seems to be rare in biological control. One well documented example, however, shows how other cases could arise in the future and illustrates how important it is in biological control to select the appropriate genetic strain of natural enemy.

The larch saw-fly, *Pristiphora erichsonii* (Fig. 9.4A), has been a serious pest of larch trees in Canada since the last century. During 1910 and 1911 an ichneumonid parasite, *Mesoleius tenthredinis*, was imported from England into Manitoba in an attempt to control the saw-fly biologically. This introduction was very successful and in the following years up to 88 per cent parasitism was recorded. During the 1940's, however, saw-fly outbreaks started to become more frequent. Muldrew (1953) showed that in Manitoba many parasitized saw-fly larvae survived because the parasite eggs were encapsulated and killed by their hosts. This resistant strain of hosts was, of course, at a considerable selective advantage and was soon to be found right across the continent from north-eastern British Columbia to Nova Scotia. Recent work suggests that host resistance may be overcome by importing a 'Bavarian strain' of the parasite which is immune to the haemocytic reaction of the saw-fly and whose progeny are also immune even when crossed with the susceptable 'Canadian strain' (Turnock & Muldrew 1971, Turnock 1972).

9.4.5 *Searching characteristics*

There have been numerous, but usually unsuccessful, biological control programmes using species of the genus *Trichogramma*, which parasitize eggs of a very wide range of Lepidoptera (Fig. 9.4B). They have been used in attempts to control several species, particularly the sugar cane moth borers in America and the West Indies (Metcalfe & Brenière 1969). Regular mass liberations of *Trichogramma* were first attempted in 1921 in Guyana and were then used more widely with parasites reared on the eggs of grain moths such as *Sitotroga cerealella*. We can only guess the reasons for the failure of *Trichogramma* in biological control. Perhaps failure of synchronization and the wide host range prevents very high parasite populations being maintained in sugar plantations, or perhaps the culturing method has selected a laboratory race especially adapted to *Sitotroga*.

It is also quite possible, however, that the basic searching characteristics of *Trichogramma* prevent it being a suitable biological control agent. Evolution does not necessarily select for a parasite that will keep its host populations at very low levels. Parasites that are always abundant year after year must have an abundant supply of hosts to maintain these high numbers. Such parasites are likely to have low searching efficiencies. All host–parasite theories so far developed are at least agreed on this: provided the parasite doesn't itself suffer some large specific mortality, the higher the searching efficiency, the lower the average densities of both host and parasite populations. The more promising parasites for biological control are thus likely to be those with high searching efficiency that normally keep their hosts at very low densities and are themselves relatively scarce.

9.5   Models and biological control

At the beginning of this chapter we outlined the typical sequence of events following the successful establishment of one or more exotic natural enemies to control a particular pest. We shall now consider whether the features of parasite behaviour described in Chapter 4 can provide a useful theoretical basis for biological control. While each of the theoretical models described probably provides a satisfactory description of particular stages in a host–parasite interaction, none is able to describe the complete sequence of events leading to successful biological control.

The models of Thompson (1924, 1930) assume that the average numbers of eggs laid per parasite is constant for a particular parasite species (i.e. independent of both host and parasite densities). This may be approximately true when few parasites are faced with a super-abundance of hosts in the initial stages following introduction. Under these conditions the models may adequately predict the initial rate of increase of the parasites but are unsatisfactory since they also predict an ultimate extinction of both host and parasite populations—which is in conflict with observation.

The theory of Nicholson (1933) (equation 4.3) is also unsuitable because the searching efficiency is assumed to be a constant for a given parasite species. Holling (1959) has shown that the searching efficiency of a parasite or predator must depend on the density of hosts, and Hassell & Varley (1969) and Hassell & Rogers (1972) have shown how searching efficiency may depend on parasite density. Because of the constant searching efficiency Nicholson's models are unstable and do not allow two or more specific, synchronized parasites to co-exist on a single host species (Varley & Gradwell 1970). We know from field observations that host–parasite interactions are far more stable than Nicholson thought and that several parasite species often co-exist on a single host species.

The quest theory of Hassell & Varley (1969) is based on experimental observations that the searching efficiency of parasites decreases as parasite density increases, largely because interference increases time-wasting activities and dispersal. This parasite interference, represented by m in equations (4.10) to (4.12) is observed from one place simply as a fall in the apparent value of the area of discovery as the parasite population rises. In parasite behaviour it manifests itself as emigration. Successful introductions for biological control spread spontaneously. Townes (1972) estimates that the ichneumonid *Pycnocryptus director*,

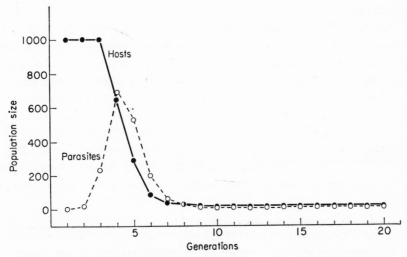

Fig. 9.5 A mathematical model simulating a successful introduction of a parasite for the biological control of a pest. The formulae used in the calculation are developed from an expansion of formula (4.5) and include provision for parasite interference and handling time.

$$N_{n+1} = F N_n \exp\left(-\frac{Na}{N_n}\right)$$

where the number of hosts attacked,

$$N_a = \frac{a' T N_n P_n^{1-m}}{1 + a' T_h N_n P_n^{-m}}$$

and the adult parasites of the next generation P_{n+1}

$$P_{n+1} = N_n\left[1 - \exp\left(-\frac{N_a}{N_n}\right)\right]$$

For the model, the rate of increase $F = 2$, the attack coefficient $a' = 0.005$, total time $T = 100$, handling time $T_h = 5$ and the mutual interference constant $m = 0.8$.

after a few years of increase in the initial area, spread at a rate of about 150 miles a year.

Parasite interference tends to stabilize interactions and permits more than one parasite species to co-exist on a given host species. The model, however, has two faults: first the observed increase in searching efficiency as parasite density falls cannot be extrapolated indefinitely; secondly, parasites must sometimes be significantly limited by egg supply or the effects of handling time, so searching efficiency cannot be entirely independent of host density. Such a model therefore predicts too rapid an increase in the parasite population following an introduction.

We have found that it is only by combining some of the features of different models that the sequence of events in biological control may be satsfactorily described. Figure 9.5 shows the outcome of a model which begins with the introduction of a few parasites to control a pest population. Such a model broadly describes the type of outcome hoped for from a successful biological control project. The searching efficiency of the parasites depends on both the host density (by the inclusion of a handling time) and parasite density (by the inclusion of parasite interference. Including a handling time tends to limit the attack rate when hosts are very abundant. This reduces the initial rate of increase of

Table 9.2 Some searching charcteristics of insect natural enemies which affect stability and average population levels. After Hassell & May (1973).

Searching characteristic	Optimum for biological control	Effect
Intrinsic searching efficiency	High	Reduced average population levels
Handling time (as a proportion of total time)	Low	Negligible reduction in stability and only very slight increase in av. pop. levels
Interference	The mutual interference constant (m) between 0 and 1	Increase in stability Some increase in average population levels
Aggregation	High	Increase in stability (depends also on host distribution). Some increase in average population levels

the parasite population. The rapid stabilization of the populations at low levels depends on the marked degree of interference included in the model. More sophisticated models (Hassell & May 1973) suggest that there are at least four searching characteristics that should be optimized in selecting natural enemies for biological control. These are listed in Table 9.2. They include characteristics which will tend to reduce the

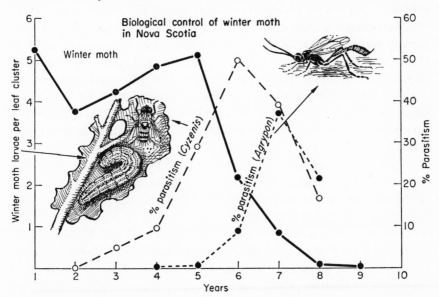

Fig. 9.6 After the establishment in Nova Scotia of the introduced parasites *Cyzenis* and *Agrypon* the winter moth population declined as the percentage of parasitism increased (Embree 1966). The graphs are average figures for seven localities. The time scale for each place was counted from the time *Cyzenis* was first observed there. For most places year 1 was 1954. Inset: adult *Cyzenis albicans* and a winter moth larva; right: adult of the introduced ichneumonid *Agrypon flaveolatum*.

average population density of the host (principally the intrinsic searching efficiency) and others which promote stability, such as a tendency to aggregate where host density is highest and for intereference between the searching natural enemies to modify their behaviour.

Although such models may give us insight into the basic searching characteristics required of a parasitic for successful biological control, they do not predict the outcome of a particular project. How much specific information do we require for a predictive model? Let us take the case of the winter moth in Canada as an example, since we have more information about the winter moth and its parasites from both Europe and Canada than is available for any other biological control experiment. After its accidental introduction into eastern Canada, the winter moth rapidly became a serious defoliator of hardwood forest trees, shade trees and orchards. A biological control programme was initiated in 1954 involving two larval parasites, a tachinid (*Cyzenis albicans*), and an ichneumonid (*Agrypon flaveolatum*). Both parasites rapidly spread from their release sites and greatly increased in numbers (Embree 1965, 1966, 1971). Figure 9.6 shows how the winter moth populations declined within six years of the appearance of *Cyzenis*. *Agrypon* appeared later than *Cyzenis* and caused high levels of parasitism only after the initial decline in the winter moth. There have been no reports of serious outbreaks of winter moth since its dramatic 'crash' following the introduction of *Cyzenis*.

On the basis of their winter moth studies in England, Varley & Gradwell (1968) predicted that in Canada there would be 'periodic outbreaks [of winter moth] causing defoliation at nine or 10-year intervals'. This prediction was based on a model in which *Cyzenis* was assumed to act in a Nicholsonian way. A new model which attempts to describe the Canadian situation rather than that found in England gives a different prediction. This model is shown in Fig. 9.7 and is based on data from Embree (1965) and from Varley & Gradwell (1968, 1971b). The model is based on the following assumptions:

1 each female winter moth lays an average of 89 eggs (F = 89) as found by Embree;
2 there is an equal sex ratio;
3 there is a maximum winter moth population above which there can be no increase;
4 the winter moth population suffers four major mortalities, k_1 to k_4.

k_1 Egg-larval mortality excluding parasitism. This is assumed to be constant ($k_1 = 1 \cdot 6$) equal to the mean of the observed values from Embree (1965).
k_2 Mortality due to *Cyzenis*. This is estimated from the submodel for parasitism by *Cyzenis* described in Section 7.9 [$k_2 = (0 \cdot 056 P^{1-0 \cdot 52})/(2 \cdot 3)$] see formula (4.12). The handling time of *Cyzenis* is extremely short and egg supply very large. The model therefore does not include any effects of host density on searching efficiency.

k_3 Mortality due to *Agrypon*. There is little known about *Agrypon* behaviour, so a submodel to describe parasitism by *Agrypon* cannot be based on measurement. We have therefore arbitrarily assumed that it acts in the same way as *Cyzenis* but has an egg limit of 100 eggs per female.

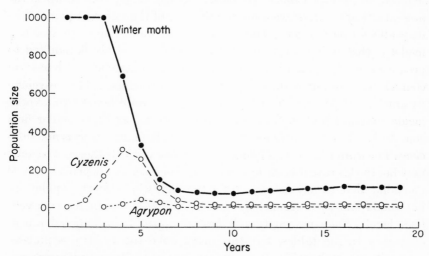

Fig. 9.7 Host–parasite model simulating the biological control of winter moth by *Cyzenis* and *Agrypon* in Nova Scotia.

Winter moth:

$$\log N_{n+1} = \log N_n - k_2 - k_3 - k_4 - k_1 + \log F$$

Cyzenis:

$$\log P_{n+1} = \log [N_n(1 - \exp{(-QP_n^{1-m})})] - k_4$$

Agrypon:

$$\log P_{n+1} = \log [N_s (1 - \exp(-QP_n^{1-m}))] - k_4.$$

(N_s = survivors from parasitism by *Cyzenis*; other symbols as previously. The values for different constants are given in the text.)

k_4 Pupal mortality. The information in Embree (1965) suggests that the pupal mortality of the winter moth in Nova Scotia is much less severe than the strongly density dependent mortality in the population studied by Varley & Gradwell. Between 1954 and 1959 the pupal mortality was represented by a k-value of about 0·2 or less. We have assumed a constant value of 0·2 in this model.

The only other components in the model concern the parasites. Since they are equally susceptible to mortality from predators whilst in the soil during their later larval and pupal stages, they are assumed to suffer the same mortality as the winter moth pupae.

We can see from Fig. 9.7 that the model predicts that the introduction of the two parasites will lead to a stable interaction with a greatly reduced mean winter moth population, as has been observed up to the present. Clearly, more accurate descriptions of the different parameters may affect this outcome. Perhaps the most important aspect of this model is that it indicates how much preliminary study is required to produce a model which is likely to predict the outcome of a biological control programme with a fair degree of confidence. The searching parameters of parasites can be obtained by careful laboratory experiments, but it is not known how these relate to searching under field conditions. An adequate sub-model for parasitism, however, is not sufficient. The winter moth example has shown how critical other components may be. In this case it is the absence of a strong density dependent pupal mortality that largely accounted for the high winter moth populations in Canada and then permitted *Cyzenis* populations to increase to very high levels. Such information is best obtained by a detailed census converted to life tables, but this may prove too lengthy a process, especially when there is only one generation of the host insect per year. Only if the factors affecting the survival of both host and parasite in the field have been adequately quantified can we develop satisfactory predictive models for use in biological control.

9.6　Some recent trends in biological control

Table 9.1 indicates that biological control has been much less successful against pests of field and garden crops than against pests of forest, fruit or ornamental trees. Only 12 out of 42 pests of field or garden crops (approximately 5 per cent) had been completely 'controlled' by natural enemies in the USA up to 1950 despite 20 beneficial species being established. This is compared with 36 per cent of orchard pests and 25 per cent of forest and ornamental tree pests that were biologically controlled in the same period. The many successes with orchard pests have been largely concerned with various scale insects. They make up nearly half of all biological control projects to date in which there has been some degree of success (DeBach *et al.* 1971).

The main hindrance to the biological control of pests of annual field crops arises from cultural practices which cause a succession of ecological upheavals in an artificially simplified environment. Successful biological control is always associated with a close interaction between pest and natural enemies. All the cases cited in this chapter have involved parasites or predators which are specific or at least show a strong host preference, and there has always been a continuous reproductive interaction between host and parasite populations.

In the annual field crop situation there are regular sequences of planting, growing, harvesting and ploughing. This often means that populations of pests and their parasites or other natural enemies start the growing season at very low densities, having died out or emigrated between harvest and the appearance of the new crop. A new population may be the survivors from this adverse period, or may be immigrants from surrounding areas, or both. Typically, the initial population of the pest suffers little mortality from natural enemies and in one or two quick generations may rapidly increase in numbers. The population of a natural enemy reaches a high level only after that of its host or prey is large and, therefore, causes really high mortalities only after the host or prey has reached pest status.

Recent work suggests that under these conditions biological control depends on establishing high levels of parasitism or predation at the start of the growing season. The work of Hussey & Parr (1963) shows one way in which this can be achieved. They found that the red spider mite (Fig. 9.4c) which damages cucumbers in glass houses could be effectively controlled by a predatory mite, *Phytoseiulus persimilis* if the spider mites 'are deliberately introduced on to every cucumber plant within a few days of planting by placing a small section of bean leaf bearing 10–20 female mites on each plant. The prey mites are allowed to increase until the mean leaf is damaged 0·4 . . . when two predators are introduced on every second plant' (Hussey & Bravenboer 1971). Thus, it is only by increasing prey populations at the beginning of the season and ensuring that they are more-or-less evenly distributed that the introduced predators can increase sufficiently rapidly to ensure an efficient interaction throughout the summer without further manipulation.

A similar technique has been developed by Parker (1971) to control the small white butterfly *Pieris rapae*, whose caterpillar is a pest of *Brassica* crops (Fig. 9.8A). Parker found that spring populations of *Apanteles glomeratus*, a braconid parasite of *Pieris*, are first active approximately two weeks before its hosts are available, and that in the first two generations *Pieris* is too scarce for the parasites to increase sufficiently to suppress subsequent *Pieris* populations in the same season. By releasing both hosts and parasites early in each season effective *Pieris* control was achieved throughout the season.

These examples of biological control show how important it may be to investigate ways of improving the efficiency of indigenous or introduced parasites once they are established. The provision of extra food for adult natural enemies can help their populations to increase more rapidly. Hagen *et al.* (1970) demonstrated very clearly that spraying artificial food for predators such as syrphids (hover flies), coccinellids (ladybird beetles), and chrysopids (lacewing flies) increased their effectiveness against aphids on alfalfa. These nutrient sprays not only attracted predators from surrounding areas, but also tended to increase their fecundity.

Table 9.3   After Schlinger & Dietrick (1960)

Types of natural enemies	Densities of natural enemies per m² of cultivated alfalfa	
	Regular farming	Strip farming
Coccinellid adults	11	51
Coccinellid larvae	3	57
Green lacewing larvae	48	51
Aphid parasites	17	71
Big-eyed bugs	49	99
Aphid-eating spiders	26	270
Totals	154	599

Modified cultural practices may be an important way of preventing the large-scale mortality or emigration of beneficial insects that often occurs when alfalfa is harvested. The spotted alfalfa aphid, *Therioaphis maculata* is an important alfalfa pest in California. This aphid is attacked by a wide range of natural enemies whose effectiveness can be increased by carefully planning the spacing and cutting of the alfalfa plants. Table 9.3 shows how the cutting of alfalfa in alternate strips increases the stock of natural enemies present (Schlinger & Dietrick 1960).

We have tried to emphasize in this chapter that classical biological control can be a very effective means of pest control under certain conditions. Failures have occurred for so many different reasons that the introduction of a particular parasite will provide the only real test of its effectiveness. We do not agree with Turnbull & Chant (1961) and Turnbull (1967) that multiple introductions of natural enemies may jeopardize biological control. Their view is supported neither on theoretical grounds (see Chapter 4) nor by practical experience. Host–parasite models have now reached the stage where they can indicate which searching characteristics are likely to result in stable and greatly reduced populations. We hope such models will help in selecting the

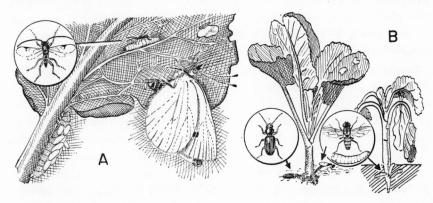

Fig. 9.8
A The small white butterfly *Pieris rapae* (rt) laying an egg on a brassica leaf. Above it two small caterpillars; one being parasitized by *Apanteles glomeratus* (enlarged, left).
Below, left, cocoons of *Apanteles* on a large dying caterpillar.
B Healthy and wilted brassica seedlings. A cabbage root-fly is laying eggs (left) and larvae are on the roots of the wilted plant (right). Inset: *Bembidion lampros* and *Erioischia brassicae* adult and larva.

most suitable parasites (as far as their searching ability is concerned) for particular biological control programmes. The current work on the manipulation of indigenous natural enemy populations suggests that this type of biological control is also likely to become increasingly important, especially in integrated projects.

Indigenous populations of parasites and predators could often be a more important cause of pest mortality if not inhibited by the insecticides used in an attempt to obtain 'chemical control'. Approximately one in every 10 species of animal is a parasitic insect. This figure includes true parasites as well as parasitoids but does not include the large numbers of predatory species. We are probably quite near the mark if we say that almost half of all insect species feed on other insects. We mention these crude estimates to help emphasize how extremely important natural enemies are likely to be as mortality factors helping to maintain other insect species at relatively low levels of abundance. The best evidence of this is indirect and comes from the many documented cases where the use of insecticides has led to pest resurgence or the appearance of new pests by elimination of indigenous natural enemies. A good example is given in Fig. 9.9A which shows the results of treating cauliflowers against cabbage root-fly (*Erioischia brassicae*) by surrounding the plot with a straw rope soaked with DDT (Wright, Hughes & Worral 1960). The percentage of plants wilting due to cabbage root-fly attack is higher than in a control plot with no treatment. Figure 9.9B strongly suggests that this is due to the devastating effect of the insecticide treatment on the carabid beetle (*Bembidion lampros*) which eats many cabbage root-fly eggs (Fig. 9.8B). This is an example of pest increase after an inappropriate insecticide treatment. There are also many cases where destruction of natural enemies has led to the appearance of new pest species. Thus Conway (1972) showed that the appearance of new pests of cocoa in Malaysia was correlated with periods of insecticide use. Wood (1968, 1971) describes how several serious outbreaks of defoliators of oil palm in Malaysia have been caused by the application of persistent insecticides against pests of minor importance. The 'cure' proved worse than the disease. Smith (1969) draws much the same picture for cotton in the Canete Valley of Peru. In both areas the pest problems have been largely overcome by applying pesticides in such a way as to minimize their effect on the natural enemy population.

Before the middle of the present century the belief was widespread that insecticides would be the most effective long-term answer to insect pest problems, but gradually there has been a change of view (Mellanby 1967). There are four main reasons.

1 Many major pests have evolved insecticide resistance to a spectrum of compounds to some of which the insect had never been subjected.

2 The increased insecticide doses needed for a satisfactory kill are costly and may lead directly to crop damage or to residue problems in the marketable crop.

3 Persistent insecticides, like the chlorinated hydrocarbons (e.g., DDT), are found to be concentrated in animals at the ends of food chains in places far removed from the sites where insecticides have been used. Chlorinated hydrocarbons have been detected in freshwater animals. in oceanic fishes and even in birds and seals on the antarctic continent.

The first analytical methods used left some doubt as to the nature and origin of the chlorinated compounds found—whether from insecticides or from similar chlorinated compounds used in paints etc.

4 The proportion of insecticide used in normal spray programmes which actually enters the bodies of the target species is extremely small.

New and more specific methods of dealing with the problem have been urgently investigated. Some of these are: *specific sex attractants*

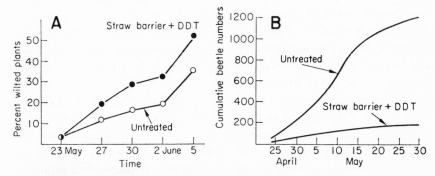

Fig. 9.9 The effect of surrounding cauliflower plants by a straw barrier soaked in DDT.

A In the treated plot a larger percentage of plants wilt from the attack of larvae of the cabbage root-fly, *Erioischia*.

B Fewer of the ground beetle *Bembidion*, which eat cabbage root-fly eggs, are found in pitfall traps within the treated plot. Data from Wright *et al.* (1960).

which lure the insects to traps where they can be killed by small amounts of insecticide; *repellants* which prevent female insects from laying eggs on the crop or make the crop unacceptable to the larvae; *synthetic hormones*—analogues of the juvenile and moulting hormones of insects which interfere with normal development of the insect, and chemicals or irradiation with X or gamma rays which can be used as *sterilants*.

The sterile male technique has had a number of successes. The screw worm is the maggot stage of the fly *Cochliomyia hominivorax* which lays its eggs in minor wounds on cattle (or man) and its effect is very like that of its relative the sheep blow-fly on sheep (Chapters 2 and 3). It was estimated that the economic losses from this pest in the southern states of the USA, amounted to some 120 million dollars per annum. As a pilot experiment, screw worm flies were artificially bred in large numbers and the males were sterilized by irradiation and then liberated on the small Caribbean island of Curaçao. The sterile males had to outnumber the wild males and mate with the wild females whose eggs would then be infertile. In a few generations the experiment completely eradicated the pest from the island. The same method was then used in the USA, where, at a cost of 60 million dollars, the screw worm was eradicated from the southern states. Because the fly still lived in Mexico, the liberation of sterile males was continued in the region of the USA–Mexican border to prevent re-invasion. The recurrent cost of this 'sanitation barrier' is some 6 million dollars per annum. Reports (Anon. 1973) suggest that some flies have penetrated the barrier, but even if much of the effort has to be repeated, the economic returns from this technique with the screw worm are considerable.

All such methods share with insecticides one important limitation on their usefulness. Unless the pest can be completely eradicated as the screw worm was on Curaçao, it may return to its original level very rapidly after any reduction in numbers (Varley 1973); the rate of recovery depends on how powerful were the density dependent influences which originally regulated its population (Chapter 2). To avoid this recovery the artificial restraints must operate in every generation, and replace the mortality factors which no longer operate. To give an example based on the sterile male technique: consider a population model in which, at low population densities, the pest can increase 10-fold. Then, to reduce the population the sterile males will have to outnumber the wild fertile males by at least 9 : 1, and more than this if there are multiple matings. Unless the pest concerned is cheap and easy to rear in quantity, the enormous rearing programme may be too costly.

Clearly the advantage of biological control of the conventional kind is that the only expense is the initial one of careful testing and of the fairly small initial introduction. Thereafter the insects do all the work, for which there is no recurrent charge.

If, in the future, pest management is to be put on a sound scientific basis, the biological system with which we wish to interfere must be well enough understood for us to model it mathematically so that we can predict which method will have the most satisfactory outcome. If this book helps people to find ways of doing this, so that we can manage populations intelligently, it will have served a useful purpose.

REFERENCES

Numbers below each reference give the section where that reference is mentioned.

ANDREWARTHA H.G. & BIRCH L.C. (1954) *The distribution and abundance of animals*. University of Chicago Press, Chicago.
2.2, 5.5

Anon (1973) The screw worm strikes back. *Nature, Lond.*, **242**, 493–4.
9.6

AUER C. (1968) Erste Ergebnisse einfacher stochastischer Modelluntersuchungen über die Ursachen der Populationsbewegung des grauen Lärchenwicklers *Zeiraphera diniana*, Gn. (= *Z. griseana* Hb.) im Oberengadin, 1949/66. *Z. angew. Ent.*, **62**, 202–235.
8.3, 8.7

AYALA F.J. (1969). Experimental invalidation of the principle of competitive exclusion. *Nature, Lond.*, **224**, 1076–1079.
3.3, 3.5

AYALA F.J. (1970) Invalidation of principle of competitive exclusion defended. *Nature, Lond.*, **227**, 89–90.
3.3, 3.5

BAKKER K. (1961) An analysis of factors which determine success in competition for food among larvae of *Drosophila melanogaster*. *Archs néerl. Zool.*, **14**, 200–281.
2.2, 2.6

BAKKER K. (1964) Backgrounds of controversies about population theories and their terminologies. *Z. angew. Ent.*, **53**, 187–208.
5.5

BAKKER K. (1969) Selection for rate of growth and its influence on competitive ability of larvae of *Drosophila melanogaster* (Dipt., Drosophilidae). *Neth. J. Zool.*, **19**, 541–595.
2.2

BALTENSWEILER W. (1964) *Zeiraphera griseana* Hübner (Lepidoptera : Tortricidae) in the European Alps. A contribution to the problem of cycles. *Can. Ent.*, **96**, 792–800.
8.3

BALTENSWEILER W. (1968) The cyclic population dynamics of the grey larch tortrix, *Zeiraphera griseana* Hübner (= *Semasia diniana* Guenée) (Lepidoptera : Tortricidae). [In] Southwood, T.R.E. (Ed.) *Insect Abundance. Symp R. ent. Soc. Lond.*, **4**, 88–97.
8.3

BARON S. (1972) *The desert locust*. Eyre Methuen, London.
5.5

BARTLETT M.S. (1949) Fitting a straight line when both variables are subject to error. *Biometrics* **5**, 207–212.
Exercise 8.2.5

BEAVER R.A. (1966) The development and expression of population tables for the bark beetle *Scolytus scolytus* (F.). *J. Anim. Ecol.*, **35**, 27–41.
8.5

BESS H.A., BOSCH R. VAN DEN & HARAMOTO F.H. (1961) Fruit fly parasites and their activities in Hawaii. *Proc. Hawaii. ent. Soc.*, **17**, 367–378.
3.8

BIRCH L.C. (1953) Experimental background to the study of the distribution and abundance of insects. I. The influence of temperature, moisture and food on the innate capacity for increase of three grain beetles. *Ecology* **34**, 608–711.
2.3

BIRCH L.C. (1957) The role of weather in determining the distribution and abundance of animals. *Cold Spring Harb. Symp. quant. Biol.*, **22**, 203–218.
5.3

BIRCH L.C., PARK T. & FRANK M.B. (1951) The effect of intraspecies and interspecies competition on the fecundity of two species of flour beetles. *Evolution, Lancaster, Pa.*, **5**, 116–132.
3.2

BLAIS J.R. (1965) Spruce budworm outbreaks in the past three centuries in the Laurentide Park, Quebec. *Forest Sci.*, **11**, 130–138.
8.3

BLANK T.H., SOUTHWOOD T.R.E. & CROSS D.J. (1967) The ecology of the partridge. I. Outline of population processes with particular reference to chick mortality and nest density. *J. Anim. Ecol.*, **36**, 549–556.
7.2

BROADHEAD E. & WAPSHERE A.J. (1966) *Mesopsocus* populations on larch in England—the distribution and dynamics of two closely-related coexisting species of Psocoptera sharing the same food resource. *Ecol. Monogr.*, **36**, 327–388.
3.6

BUCKNER C.H. (1969) The common shrew (*Sorex araneus*) as a predator of the winter moth (*Operophtera brumata*) near Oxford, England. *Can. Ent.*, **101**, 370–375.
7.7, 8.7

BURNETT T. (1956) Effects of natural temperatures on oviposition of various numbers of an insect parasite (Hymenoptera, Chalcididae, Tenthredinidae). *Ann. ent. Soc. Am.*, **49**, 55–59.
4.4–4.6

BURNETT T. (1958) A model of host–parasite interaction. *Proc. 10th Int. Congr. Ent.*, **2**, 679–686.
4.4, 4.6

BUXTON P.A. & LEWIS D.J. (1934) Climate and tsetse flies: laboratory studies upon *Glossina submorsitans* and *tachinoides*. *Phil. Trans. R. Soc.*, (B) **224**, 175–240.
5.4.4

CAMPBELL R.W. (1967) The analysis of numerical change in gypsy moth populations. *Forest Sci. Monogr.*, **15**, 1–33.
8.4

CANNING E.U. (1960) Two new microsporidian parasites of the winter moth, *Operophtera brumata* (L.). *J. Parasit.*, **46**, 755–763.
7.4.5

CHENG L. (1970) Timing of attack by *Lypha dubia* Fall. (Diptera: Tachinidae) on the winter moth *Operophtera brumata* (L.) (Lepidoptera: Geometridae) as a factor affecting parasite success. *J. Anim. Ecol.*, **39**, 313–320.
7.8

CLARK L.R., GEIER P.W., HUGHES R.D. & MORRIS R.F. (1967) *The ecology of insect populations in theory and practice*. Methuen, London.
8.2

CLAUSEN C.P. (1956) Biological control of insect pests in the continental United States. *Tech. Bull. U.S. Dep. Agric.*, **1139**, i–vi, 1–151.
9.4

CONWAY G. (1972) Ecological aspects of pest control in Malaysia. [In] Farrar M.T. & Milton J.P. (Eds.) *Careless Technology: Ecology and International Development*.
9.6

COOK L.M. (1971) *Coefficients of natural selection*. Hutchinson University Library, London. pp. 207.

COOPE G.R. (1970) Interpretations of quaternary insect fossils. *A. Rev. Ent.*, **15**, 97–120.
5.3

CORBET P.S. (1956) Environmental factors influencing the induction and termination of diapause in the Emperor Dragonfly, *Anax imperator* Leach (Odonata: Aeshnidae). *J. exp. Biol.*, **33**, 1–14.
5.4.1

CORBET P.S. (1962) *A biology of dragonflies*. Witherby Ltd., London.
5.4.3

CROMBIE A.C. (1945) On competition between different species of graminivorous insects. *Proc. R. Soc.* (*B*), **132**, 362–395.
2.2, 2.3, 3.4

CROMBIE A.C. (1946). Further experiments on insect competition. *Proc. R. Soc.* (*B*), **133**, 76–109.
3.3, 3.4

DAVIDSON J. (1944) On the relationship between temperature and rate of development of insects at constant temperatures. *J. Anim. Ecol.*, **13**, 26–38.
5.4.3

DAVIDSON J. & ANDREWARTHA H.G. (1948a) Annual trends in a natural population of *Thrips imaginis* (Thysanoptera). *J. Anim. Ecol.*, **17**, 193–199.
5.5, 8.7

DAVIDSON J. & ANDREWARTHA H.G. (1948b) The influence of rainfall, evaporation and atmospheric temperature on fluctuations in the size of a natural population of *Thrips imaginis* (Thysanoptera). *J. Anim. Ecol.*, **17**, 200–222.
5.5, 8.7

DeBach P. (1965) Weather and the success of parasites in population regulation. *Can. Ent.*, **97**, 848–863.
9.4.1

DeBach P. (1971) The use of imported natural enemies in insect pest management ecology. *Proc. Tall Timbers Conf. on Ecol. Anim. Control by Habitat Management*. No. 3, 1971, pp. 211–233. Tallahassee, Florida.
9.4

DeBach P. & Smith H.S. (1941) The effect of host density on the rate of reproduction of entomophagous parasites. *J. econ. Ent.*, **34**, 741–745.
4.4, Exercise 4.2

DeBach P., Rosen D. & Kennett C.E. (1971) Biological control of coccids by introduced natural enemies. [In] Huffaker C.B. (Ed.) *Biological Control*, pp. 165–194. Plenum Press, New York.
9.3, 9.4, 9.6

DeBach P. & Sundby P.A. (1963) Competitive displacement between ecological homologues. *Hilgardia* **34**, 105–166.
3.8

Dixon A.F.G. (1958) The escape responses shown by certain aphids to the presence of the coccinellid beetle *Adalia decempunctata* (L.). *Trans. R. ent. Soc. Lond.*, **110**, 319–334.
4.7

Doutt R.L. (1954) An evaluation of some natural enemies of the olive scale. *J. econ. Ent.*, **47**, 39–43.
9.4.2

Duncan D.P. & Hodson A.C. (1958) Influence of the forest tent caterpillar upon the Aspen forests of Minnesota. *Forest Sci.*, **4**, 71–93.
8.3

Embree D.G. (1965) The population dynamics of the winter moth in Nova Scotia, 1954–1962. *Mem. ent. Soc. Can.*, **46**, 1–57.
9.5

Embree D.G. (1966) The role of introduced parasites in the control of the winter moth in Nova Scotia. *Can. Ent.*, **98**, 1159–1168.
9.5

Embree D.G. (1971) The biological control of the winter moth in eastern Canada by introduced parasites. [In] Huffaker C.B. (Ed.) *Biological Control*, pp. 217–226. Plenum Press, New York.
9.4, 9.5

Frank J.H. (1967) The effect of pupal predators on a population of winter moth, *Operophtera brumata* (L.) (Hydriomenidae). *J. Anim. Ecol.*, **36**, 611–621.
7.7, 8.7

Fuller M.E. (1934) The insect inhabitants of carrion: a study in animal ecology. *Bull. Coun. scient. ind. Res., Melb.*, **82**, 1–62.
3.7

Gause G.F. (1934) *The struggle for existence*. Hafner, New York (reprinted 1964).
3.3

GAUSE G.F. & WITT A.A. (1935) Behavior of mixed populations and the problem of natural selection. *Am. Nat.*, **69**, 596–609.
3.4

GILPIN M.E. & JUSTICE K.E. (1972) Reinterpretation of the invalidation of the principle of competitive exclusion. *Nature, Lond.*, **236**, 273–301.
3.5

HAGEN K.S., SAWALL E.F. & TASSAN R.L. (1970) The use of food sprays to increase effectiveness of entomophagous insects. *Proc. Tall Timbers Conf. on Ecological Animal Control by Habitat Management*, **2**, 59–81.
9.6

HARDWICK D.F. (1971) The 'phenological date' as an indicator of the flight period of noctuid moths. *Can. Ent.*, **103**, 1207–1216.
5.4.3

HASSELL M.P. (1966) Evaluation of parasite or predator responses. *J. Anim. Ecol.*, **35**, 65–75.
4.8

HASSELL M.P. (1968) The behavioural response of a tachinid fly (*Cyzenis albicans* (Fall.)) to its host, the winter moth (*Operophtera brumata* (L.)). *J. Anim. Ecol.*, **37**, 627–639.
7.4.3

HASSELL M.P. (1969a) A study of the mortality factors acting upon *Cyzenis albicans* (Fall.), a tachinid parasite of the winter moth *Operophtera brumata* (L.)). *J. Anim. Ecol.*, **38**, 329–339.
7.8

HASSELL M.P. (1969b) A population model for the interaction between *Cyzenis albicans* (Fall.) (Tachinidae) and *Operophtera brumata* (L.) (Geometridae) at Wytham, Berkshire. *J. Anim. Ecol.*, **38**, 567–576.

HASSELL M.P. (1971a) Mutual interference between searching insect parasites. *J. Anim. Ecol.*, **40**, 473–486.
4.6

HASSELL M.P. (1971b) Parasite behaviour as a factor contributing to the stability of insect host–parasite interactions. [In] Boer, P.J. den & Gradwell G.R. (Eds.) *Dynamics of populations. Proc. Adv. Study Inst. Dynamics Numbers Popul.* Osterbeek, 1970, pp. 366–379. Wageningen.
4.6

HASSELL M.P. & MAY R. (1973) Stability in insect host–parasite models. *J. Anim. Ecol.*, **42**, 693–726.
9.5

HASSELL M.P. & ROGERS D.J. (1972) Insect parasite responses in the development of population models. *J. Anim. Ecol.*, **41**, 661–676.
4.5, 4.8, 9.5

HASSELL M.P. & VARLEY G.C. (1969) New inductive population model for insect parasites and its bearing on biological control. *Nature, Lond.*, **223**, 1133–1137.
4.6, 9.5

G

HIGGINS L.G. & RILEY N.D. (1970) *A field guide to the butterflies of Britain and Europe.* Collins, London.
5.3

HODSON A.C. (1941) An ecological study of the forest tent caterpillar, *Malacosoma disstria* Hbn., in northern Minnesota. *Tech. Bull. Minn. agric. Exp. Sta.*, **148**, 1–55.
8.3

HOLLING C.S. (1959) Some characteristics of simple types of predation and parasitism. *Can. Ent.*, **91**, 385–398.
4.5, 9.5

HOLLING C.S. (1966) The functional response of invertebrate predators to prey density. *Mem. ent. Soc. Can.*, **48**, 1–86.
4.5

HOPKINS A.D. (1918) Periodical events and natural law as a guide to agricultural research and practice. *Mon. Weath. Rev. U.S. Dep. Agric. Suppl.* **9**, 1–42.
5.4.3

HOWARD L.O. & FISKE W.F. (1911) The importation into the United States of the parasites of the gipsy-moth and the brown-tail moth. *Bull. Bur. Ent. U.S. Dep. Agric.*, **91**, 1–312.
2.4, 4.4, 5.5

HUFFAKER C.B. & KENNETT C.E. (1966) Studies of two parasites of olive scale, *Parlatoria oleae* (Colvée). IV. Biological control of *Parlatoria oleae* (Colvée) through the compensatory action of two introduced parasites. *Hilgardia*, **37**, 283–335.
9.4, 9.4.2

HUFFAKER C.B. & KENNETT C.E. (1969) Some aspects of assessing efficiency of natural enemies. *Can. Ent.*, **101**, 425–447.
9.4

HUFFAKER C.B., KENNETT C.E. & FINNEY G.L. (1962) Biological control of olive scale, *Parlatoria oleae* (Colvée), in California by imported *Aphytis maculicornis* (Masi) (Hymenoptera: Aphelinidae). *Hilgardia*, **32**, 541–636.
9.4, 9.4.2

HUSSEY N.W. & PARR W.J. (1963) The effect of glasshouse red spider mite (*Tetranychus urticae* Koch) on the yield of cucumbers. *J. hort. Sci.*, **38**, 255–263.
9.6

HUSSEY N.W. & BRAVENBOER L. (1971) Control of pests in glasshouse culture by the introduction of natural enemies. [In] Huffaker C.B. (Ed.) *Biological Control*, pp. 195–216. Plenum Press, New York.
9.6

JOHNSON C.G. (1969) *Migration and dispersal of insects by flight.* Methuen, London.
5.5

KEY K.H.L. (1945) The general ecological characteristics of the outbreak areas and outbreak years of the Australian plague locust (*Chortoicetes terminifera* Walk.). *Bull. Counc. scient. ind. Res. Melb.*, **186**, 1–127.
5.5

KLOMP H. (1962) The influence of climate and weather on the mean density level, the fluctuations and the regulation of animal populations. *Archs néerl. Zool.*, **15**, 68–109.
5.5

KLOMP H. (1966) The dynamics of a field population of the pine looper, *Bupalus piniarius* L. (Lep., Geom.). *Adv. ecol. Res.*, **3**, 207–305.
8.3

KOEBELE A. (1890) Report of a trip to Australia made under direction of the Entomologist to investigate the natural enemies of the Fluted Scale. *Bull. Bur. Ent. U.S. Dep. Agric.*, **21** (*Rev. Ed.*), 1.32.
9.3

KREBS J.R. (1970a) Territory and breeding density in the Great Tit, *Parus major* L. *Ecology*, **52**, 2–22.
7.2

KREBS J.R. (1970b) Regulation of numbers in the Great Tit (Aves: Passeriformes). *J. Zool., Lond.*, **162**, 317–333.
7.2

KUENEN D.J. (1958) Some sources of misunderstanding in the theories of regulation of animal numbers. *Archs néerl. Zool.*, **13**, Suppl. 1, 335–341.
5.5

LEES A.D. (1955) *The physiology of diapause in Arthropods.* Cambridge University Press, Cambridge.
5.4.1

LLOYD M. (1965) Laboratory studies with confined cannibalistic populations of flour beetles (*Tribolium castaneum*) in a cold-dry environment 1. Data for 24 unmanipulated populations. *Tribolium Info. Bull.*, **8**, 89–123.
2.3

LLOYD M. (1968) Self regulation of adult numbers by cannibalism in two laboratory strains of flour beetles (*Tribolium castaneum*). *Ecology*, **49**, 245–259.
2.3

LLOYD M. & DYBAS H.S. (1966) The periodical cicada problem. 1. Population ecology. *Evolution, Lancaster, Pa.*, **20**, 133–149.
5.4.3

LOTKA A.J. (1925) *Elements of physical biology.* Baltimore.
3.4

LOTKA A.J. (1931) The structure of a growing population. *Hum. Biol.*, **3**, 459–493.
3.4

LOTKA A.J. (1932) The growth of mixed populations: two species competing for a common food supply. *J. Wash. Acad. Sci.*, **22**, 461–469.
3.4

LUCK R.F. (1971) An appraisal of two methods of analysing insect life tables. *Can. Ent.*, **103**, 1261–1271.
7.7, 8.7

McNew G.L. (1971) The Boyce Thompson Institute program in forest Entomology that led to the discovery of pheromones in bark beetles: Symposium on Population Attractants. *Contr. Boyce Thompson Inst. Pl. Res.*, **24** (1970), 251–262.
8.5

Malthus T.R. (1798) *An essay on the principle of population as it effects the future improvements of society.* London. [Reprinted by Macmillan, New York].
2.3

Mellanby K. (1967) *Pesticides and pollution.* Collins, London.
9.6

Mertz D.B. & Davies R.B. (1968) Cannibalism of the pupal stage by adult flour beetles: an experiment and a stochastic model. *Biometrics*, **24**, 247–275.
3.5

Metcalfe J.R. & Brenière J. (1969) Egg parasites (*Trichogramma* spp.) for control of sugar cane moth borers. [In] *Pests of sugar cane*, pp. 81–115. Elsevier.
9.4.5

Miller R.S. (1964) Larval competition in *Drosophila melanogaster* and *D. simulans*. *Ecology*, **45**, 132–148.
3.5

Milne A. (1957) Theories of natural control of insect populations. *Cold Spring Harb. Symp. quant. Biol.*, **22**, 253–271.
2.2

Morris R.F. (1959) Single-factor analysis in population dynamics. *Ecology*, **40**, 580–588.
7.6, 8.3

Morris R.F. (1963) The dynamics of epidemic spruce budworm populations. *Mem. ent. Soc. Can.*, **31**, 1–332.
6.4, 8.3

Muldrew J.A. (1953) The natural immunity of the larch sawfly (*Pristiphora erichsonii* (Htg.)) to the introduced parasite *Mesoleius tenthredinis* Morley in Manitoba and Saskatchewan. *Can. J. Zool.*, **31**, 313–332.
9.4.4

Murdie G. & Hassell M.P. (1973) Food distribution, searching success and predator–prey models. *Proc. Symp. Math. Theory Dynamics Biol. Popul.* 1972 (ed. M. S. Bartlett & R. W. Hiorns) pp. 87–101. Academic Press, London and New York.
4.8

Nicholson A.J. (1933) The balance of animal populations. *J. Anim. Ecol.*, **2**, 132–178.
2.4, 2.6, 4.4, 5.5, 9.5

Nicholson A.J. (1954) An outline of the dynamics of animal populations. *Aust. J. Zool.*, **2**, 9–65.
2.6, 5.5

Nicholson A.J. & Bailey V.A. (1935) The balance of animal populations. *Part I.* *Proc. zool. Soc. Lond.*, 1935, 551–598.
4.4

PARK T. (1948) Experimental studies of interspecies competition. I. Competition between populations of the flour beetles, *Tribolium confusum* Duval and *Tribolium castaneum* Herbst. *Ecol. Monogr.*, **18**, 265–308.
3.3

PARK T. (1954) Experimental studies of interspecies competition. II. Temperature, humidity, and competition in two species of *Tribolium*. *Physiol. Zoöl.*, **27**, 177–238.
3.3

PARKER F.D. (1971) Management of pest populations by manipulating densities of both hosts and parasites through periodic releases. [In] Huffaker C.B. (Ed.) *Biological Control*, pp. 365–376. Plenum Press, New York.
9.6

PEARL R. & REED L.J. (1920) On the rate of growth of the population of the United States since 1790 and its mathematical representation. *Proc. nat. Acad. Sci., U.S.A.*, **6**, 275–188.
2.3

PONTIN A.J. (1969) Experimental transplantation of nest-mounds of the ant *Lasius flavus* (F.) in a habitat containing also *L. niger* (L.) and *Myrmica scabrinodis* Nyl. *J. Anim. Ecol.*, **38**, 747–754.
3.6

REDDINGIUS J. (1971) Models as research tools. [In] Boer, P.J. den & Gradwell G.R. (Eds.) *Dynamics of Populations. Proc. Adv. Study Inst. Dynamics Numbers Popul.* Oosterbeek 1970, pp. 64–76.
5.5

RICHARDS O.W. & WALOFF N. (1954) Studies on the biology and population dynamics of British grasshoppers. *Anti-Locust Bull.*, **17**, 1–182.
1.3

RICKER W.E. (1954) Stock and recruitment. *J. Fish. Res. Bd Can.*, **11**, 559–623.
2.5

ROGERS D.J. (1972) Random search and insect population models. *J. Anim. Ecol.*, **41**, 369–383.
4.5

ROYAMA T. (1971) A comparative study of models for predation and parasitism. *Researches Popul. Ecol. Kyoto Univ. Suppl.* **1**, 1–91.
4.5

SANG J.H. (1950) Population growth in *Drosophila* cultures. *Biol. Rev.*, **25**, 188–219.
2.3

SCHLINGER E.I. & DIETRICK E.J. (1960) Biological control of insect pests aided by strip-farming alfalfa in experimental program. *Calif. Agric.*, **14**, 8–9.
9.6

SCHVESTER D. (1971) *Matsucoccus feytaudi* Duc. et 'Dépérissement' du Pin maritime. [In] *La lutte biologique en forêt. Ann. Zool.-Ecol. anim.*, *Numéro hors série*, pp. 139–151 (Institut National de la Recherche Agronomique Publ. 71–3).
8.4

SCHWERDTFEGER F. (1935) Studien über den Massenwechsel einiger Forstschäd-
linge. *Z. Forst-u. Jagdw.*, **67**, 15–38, 85–104, 449–482, 513–540.
8.2

SCHWERDTFEGER F. (1941) Über die Ursachen des Massenwechsels der Insekten.
Z. angew. Ent., **28**, 254–303.
8.2

SHELFORD V.E. (1927) An experimental investigation of the relations of the codling
moth to weather and climate. *Bull. Ill. St. nat. Hist. Surv.*, **16**, 311–440.
5.4.3

SMITH F.E. (1961) Density dependence in the Australian Thrips. *Ecology*, **42**,
403–407.
5.5

SMITH H.S. (1935) The role of biotic factors in the determination of population
densities. *J. econ. Ent.*, **28**, 873–898.
2.4, 4.4

SMITH R.F. (1969) The new and the old in pest control. Accademia Nazionale dei
Lincei. Quaderno N.128—Atti del Convegno Internazionale sul tema: *Nuove
prospettive nella lotta contro gli insetti nocivi* (Roma, 16–18 settembre 1968)
pp. 21–30.
9.6

SNEDECOR G.W. & COCHRAN W.G. (1967) *Statistical methods.* (6th edition). Iowa
State Univ. Press, Ames, Iowa.
Exercise 8.2.5

SOLOMON M.E. (1949) The natural control of animal populations. *J. Anim. Ecol.*,
18, 1–35.
2.2, 4.5

SOUTHERN H.N. (1970) The natural control of a population of Tawny Owls (*Strix
aluco*). *J. Zool., Lond.*, **162**, 197–285.
7.2

SOUTHWOOD T.R.E. (1966) *Ecological methods with particular reference to the study
of insect populations.* Methuen, London.
1.3, 2.6

SPRINGETT B.P. (1968) Aspects of the relationship between burying beetles,
Necrophorus spp. and the mite, *Poecilochirus necrophori* Vitz. *J. Anim. Ecol.*,
37, 417–424.
3.7

TAYLOR T.H.C. (1937) *The biological control of an insect in Fiji.* An account of the
coconut leaf mining beetle and its parasite complex. Imperial Institute of
Entomology, London.
9.4, 9.4.3

THOMPSON G.H. & SKINNER E.R. (1960) *The alder woodwasp and its insect enemies.*
[16 mm film in colour distributed by Oxford Scientific Films Ltd., Long
Handborough, Oxon.]
8.6

THOMPSON W.R. (1924) La théorie mathématique de l'action des parasites
entomophages et le facteur du hasard. *Annls Fac. Sci. Marseille*, **2**, 69–89.
4.3, 9.5

THOMPSON W.R. (1930) The utility of mathematical methods in relation to work on biological control. *Ann. appl. Biol.*, **17**, 641–648.
9.5

TOWNES H. (1972) Ichneumonidae as biological control agents. *Proc. Tall Timbers Conf. on ecological animal control by habitat management*, **3**, 235–248.
9.5

TURNBULL A.L. (1967) Population dynamics of exotic insects. *Bull. ent. Soc. Am.*, **13**, 333–337.
9.6

TURNBULL A.L. & CHANT D.A. (1961) The practice and theory of biological control of insects in Canada. *Can. J. Zool.*, **39**, 697–753.
9.4, 9.6

TURNOCK W.J. (1972) Geographical and historical variability in population patterns and life systems of the larch sawfly (Hymenoptera: Tenthredinidae). *Can. Ent.*, **104**, 1883–1900.
9.4.4

TURNOCK W.J. (1973) Factors influencing the fall emergence of *Bessa harveyi* (Tachinidae: Diptera). *Can. Ent.*, **105**, 399–409.
5.4.1

TURNOCK W.J. & MULDREW J.A. (1971) Parasites. From: Toward Integrated Control. *U.S.D.A. Forest Service Research Paper* NE-194, 59–87.
9.4.4

ULLYETT G.C. (1947) Mortality factors in populations of *Plutella maculipennis* Curtis (Tineidae: Lep.), and their relation to the problem of control. *Ent. Mem. Dep. Agric. Un. S. Afr.*, **2**, 77–202.
5.4.2

UVAROV B.P. (1931) Insects and climate. *Trans. R. ent. Soc. Lond.*, **79**, 1–247.
5.5

VARLEY G.C. (1947) The natural control of population balance in the knapweed gall-fly (*Urophora jaceana*). *J. Anim. Ecol.*, **16**, 139–187.
4.4, 5.4.4, 6.4, 6.8, 6.9

VARLEY G.C. (1949) Population changes in German forest pests. *J. Anim. Ecol.*, **18**, 117–122.
8.2

VARLEY G.C. (1959). The biological control of agricultural pests. *J. R. Soc. Arts*, **107**, 475–490.
9.3

VARLEY G.C. (1963) The interpretation of change and stability in insect populations. *Proc. R. ent. Soc. Lond. (C)*, **27**, 52–57.
5.5

VARLEY, G.C. (1971) The effects of natural predators and parasites on winter moth populations in England. *Proc. Tall Timbers Conf. on Ecol. Anim. Control by Habitat Management*, No. 2. Tallahassee, Fla. 103–116.
7.3

VARLEY G.C. (1973) Population dynamics and pest control. [In] Price Jones D. & Solomon M.E. (Eds.) *Biology in pest and disease control*. Blackwell, Oxford.
9.6

VARLEY G.C. & EDWARDS R.L. (1957) The bearing of parasite behaviour on the dynamics of insect host and parasite populations. *J. Anim. Ecol.*, **26**, 471–477.
4.5

VARLEY G.C. & GRADWELL G.R. (1963) Predatory insects as density dependent mortality factors. *Proc. 16 Int. Congr. Zool.*, **1**, 240.
7.7

VARLEY G.C. & GRADWELL G.R. (1968) Population models for the winter moth. [In] Southwood T.R.E. (Ed.) *Insect abundance. Symp. R. ent. Soc. Lond.*, **4**, 132–142.
7.7, 8.7, 9.5

VARLEY G.C. & GRADWELL G.R. (1970) Recent advances in insect population dynamics. *A. Rev. Ent.*, **15**, 1–24.
8.3, 9.5

VARLEY G.C. & GRADWELL G.R. (1971a) Can parasites avoid competitive exclusion? *Proc. 13 Int. Congr. Ent.* 1968, **1**, 571–572.
7.9

VARLEY G.C. & GRADWELL G.R. (1971b) The use of models and life tables in assessing the role of natural enemies. [In] Huffaker C.B. (Ed.) *Biological Control*, pp. 93–112. Plenum Press, New York.
8.7, 9.5

VERHULST P.F. (1838) Notice sur le loi que la population suit dans son accroissement. *Corresp. Math. Phys.*, **10**, 113–121.
2.3

VOLTERRA V. (1926) Variations and fluctuations of the number of individuals in animal species living together. [Translation in] Chaptman R.N., 1931, *Animal Ecology*, pp. 409–448. McGraw-Hill, New York.
3.4

VOÛTE A.D. (1946) Regulation of the density of the insect-populations in virgin-forests and cultivated woods. *Archs néerl. Zool.*, **7**, 435–470.
8.3E

VOÛTE A.D. (1964) Harmonious control of forest insects. *Int. Rev. Forest. Res.*, **1**, 325–383.
8.3E

WATSON A. (1971) Key factor analysis, density dependence and population limitation in red grouse. [In] Boer P.J. den & Gradwell G.R. (Eds.) *Dynamics in populations. Proc. Adv. Study Inst. Dynamics Numbers Popul.* Oosterbeek, 1970, pp. 548–564. Wageningen.
7.2

WELLINGTON W.G. (1957) The synoptic approach to studies of insects and climate. *A. Rev. Ent.*, **2**, 143–162.
5.4

WHITE T.C.R. (1969) An index to measure weather-induced stress of trees associated with outbreaks of Psyllids in Australia. *Ecology*, **50**, 905–909.
8.7

WITTER J.A., KULMAN H.M. & HODSON A.C. (1972) Life tables for the forest tent caterpillar. *Ann. ent. Soc. Am.*, **65**, 25–31.
8.3

WOOD B.J. (1968) *Pests of oil palms in Malaysia and their control*. The Incorporated Society of Planters, Kuala Lumpur, Malaysia.
9.6

WOOD B.J. (1971) Development of integrated control programs for pests of tropical perennial crops in Malaysia. [In] Huffaker C.B. (Ed.) *Biological Control*, pp. 422–457. Plenum Press, New York.
9.6

WRIGHT D.W., HUGHES R.D. & WORRALL J. (1960) The effect of certain predators on the numbers of cabbage root fly (*Erioischia brassicae* (Bouché)) and on the subsequent damage caused by the pest. *Ann. appl. Biol.*, **48**, 756–763.
9.6

EXPERIMENTS AND EXERCISES

Chapter 1

1.1 Changes in numbers were expressed in many ways in Table 1.1. Draw a graph of the k-value on the ordinate (scaled from 0–2) against percentage mortality (scaled from 0–100 per cent) on the abscissa. Use the figures in Table 1.1 and calculate suitable intermediate points. Draw a smooth curve through them.

1.2 Suppose that successive counts of an insect during a generation were 1000 eggs, 576 first instar, 343 second instar, 172 third instar, 93 fourth instar larvae, 51 prepupae, 26 pupae and 11 adults. Express the changes as in Table 1.1.
Between which consecutive stages (a) did most die?

(b) the greatest fraction survive?

(c) the greatest fraction die?

(d) what was the k-value for (c) to nearest 2 figs?

Three figure accuracy is quite good enough for the calculations which can easily be done on a slide rule.

Chapter 2

2.1 *Experiment to demonstrate intraspecific competition.* The most suitable insects are those which can be easily cultured and have a short developmental time. We have used house-fly, or various stored product beetles. The blow-fly *Phormia* reared from the fishermans maggots, or *Drosophila*, would also be suitable. Experiments would be equally simple with freshwater crustacea or with the seeds of flowering plants. Set up a series (2, 4, 8, 16 etc) of eggs on a fixed amount of food. Analysis of results is easy if the experiment is scored when the organisms have completed their growth, but not themselves reproduced. See Fig. 2.12 and Table 3.2A, C.

2.2 Copy the line P in Fig. 2.6A on graph paper and mark in a scale from 0–100 for the population density on the abscissa. For population densities of 1, 10, 20, 30 . . . 90, 99 tabulate the percentage mortality. Convert to k-value. Plot this on the ordinate.

Question (a) How far can you see the consequences of a density dependent mortality as so defined?

2.3 Plot the corresponding reproduction curve assuming a rate of increase $F = 2$. Using the construction in Fig. 2.7C read off successive values of the population for a number of generations. Does the density dependent mortality stabilize the population under these conditions? (yes or no.)

192

2.4 Plot on the same graph the reproduction curve assuming a reproductive rate $F = 3$. Describe the nature of the population changes over a number of generations.

Question Are the populations (a) stable, (b) unstable between definable limits, (c) indeterminate, (d) heading quickly to extinction?

2.5 On the same graph plot the curve for $F = 5$. Which of the above possibilities is now correct?

2.6 Plot a logarithmic reproduction curve for the same line P in Fig. 2.6A. Put in the diagonals which correspond to the reproductive rates $F = 2$, 3 and 5, and mark in the corresponding scales on the ordinate. You should be able to find graphical constructions which define (a) the highest value of F at which the population tends to stabilize, (b) the lowest value of F at which the population will become extinct.

2.7 Draw logarithmic reproduction curves to correspond with the density dependent relations defined by curves Q, R and S in Fig. 2.6A or with any other curve of your choice. From their shape determine the limits of their ability to stabilize a population model in which they represent the sole mortality operating.

2.8 In Table 3.2 lines A and C tabulate the numbers of adult progeny from different numbers in the parental generation of the beetles *Cathartus* and *Cryptolestes*. Tabulate for each species the number of progeny divided by the initial parental population. With parental number on the abscissa plot the progeny per beetle on the ordinate. Replot the figures with the initial number on a logarithmic scale.

Question (a) What is the formula for the best straight line through the points for each species? (two figure accuracy is enough).

2.9 The table below gives the results of an experiment where different numbers of the flour moth *Ephestia cautella* emerging from counted eggs competed for 25 g of food (wheatfeed) (data kindly supplied by D.J.Rogers).

Initial number of eggs	10	20	50	100	200	400	800	1600	3200	5000
Pupae produced	7·7	15	37	74	137	279	477	392	380	321

Calculate the k-values for the mortality and plot the k-values against the log of the initial egg number. Comment on the relationship obtained.

Questions (a) Have the larvae competed by the process termed 'scramble' or 'contest'? (b) Is the slope of the line for the first 7 points best described by a slope of $b = 0·05$, $0·2$ or $0·6$? (c) For the last 4 points is $b = 0·4$, $0·8$ or $1·2$? (d) What is the minimum amount of food required for a pupa to be produced?

Chapter 3

3.1 *Experiments to demonstrate interspecific competition.* Devise and set up your own experiment with different numbers of two competing species using the plan in Table 3.2 as a basis. You can start with either eggs or adults. Alternative pairs of competing species are *Drosophila melanogaster* and *D. subobscura* which are the pale and dark fruit flies that most often come to banana bait in most parts of the

UK, *Tribolium* species as used by Park (Fig. 3.1) etc., green bottle fiies and blue bottle flies (*Lucilia* and *Calliphora*). This is a summer experiment, best done in the open air because of the smell. Very quick—can be scored as puparia or as flies. Similar experiments could be devised for freshwater crustacea for seedling competition etc. For simple analysis the experiment should be terminated when the first generation has completed its growth and can most easily be scored. More complex experiments in which one species is given access to the limited food supply before the other might be tried.

3.2 *Exercises.* Use the figures in Table 3.2 to find the effect on the mean reproductive rate of 16 *Cathartus* or 16 *Cryptolestes* of adding extra members of the same and of the other species. Find a way to graph the effects which provides a good linear relation.

Questions   (a) Do the results confirm equation (3.4)?

(b) Is the adverse effect of *Cryptolestes* on *Cathartus* bigger than the effect of *Cathartus* on *Cryptolestes*?

Chapter 4

4.1 *Experiments on searching movements.* If you have stocks of parasitic or predatory insects, set up experiments to show how they discover their host or prey. If not, then the behaviour of hungry house-flies or blow-flies searching for tiny droplets of sugar solution can be used to demonstrate the principles. The flies can be made to walk by gumming the wings together. An arena of (say) 40 cm² can be constructed with two sheets of glass separated by slats of wood. Mark on a sheet of paper 64 squares 4 × 4 cm. In the middle of 32 of the squares put one clear mark. Put two marks on each of 16 squares, on 8 four marks and on 4 a group of eight marks. Put a sheet of glass over the paper, and on the glass put a *very small* drop of sugar solution over each mark. A hungry fly must not be satisfied even by thirty such drops. Quickly put the wooden slats around the edge, put the top glass on and insert a fly. Follow its movements with a chinagraph pencil on the top glass. A timekeeper can call out time intervals which can also be marked. (If the air is very dry, put strips of moist filter paper on outer margin of the arena.)

(a) Are the fly's movements random?

(b) Is the rate of turning altered by finding food?

(c) Is the chance any one drop is found by the fly independent of the proximity of other drops?

If a fly ceases to perform, wash the glass with a damp cloth and set up fresh sugar drops for another fly.

4.2   The host and parasite densities given in Table A come from a series of experiments by DeBach & Smith (1941). The chalcid *Nasonia vitripennis* searched for 1 day for house-fly puparia mixed in a fixed volume of wheat grains. For each combination of host and parasite density calculate the hosts found per parasite and the area of discovery using a re-arrangement of formula (4.4)

$$a = \frac{2 \cdot 3}{P} \log \frac{N}{S}$$

Table A The parasite *Nasonia* searches for 1 day for puparia of the house-fly *Musca* in a fixed volume of wheat grains.

	Parasite density	Host density	
		Initial (N)	Final (S)
Experiment A	18	36	15·4
	21	31	12·9
	18	26	11·0
	15	22	11·3
	11	23	14·3
	9	29	18·3
	11	37	23·5
Experiment B	40	1	0·111
	40	5	1·00
	40	25	7·43
	40	50	18·5
	40	100	51·7
	40	200	146·0
	40	300	238·0

For each of the 14 sets of figures tabulate the number of hosts found per parasite and calculate the area of discovery.

(a) Find the nature of the functional response of the parasite to host density.

(b) What factors could account for differences in these responses and in the areas of discovery between the two experiments?

4.3 From Table B calculate the 'area of discovery' for each density of the parasite *Nemeritis canescens* searching for a constant density of the flour moth *Ephestia cautella* (data from Hassell (1971a)).

Table B The parasite *Nemeritis* searches for larvae of the flour moth *Ephestia*.

Parasite density (P)	Host density	
	Initial (N)	Final (S)
1	532	449·6
2	532	439·5
4	532	410·4
8	532	376·9
16	532	355·5
32	532	311·8

Plot graphs of (a) a against P.

(b) $\log a$ against $\log P$.

(c) What formula describes this relationship?

4.4   Use this formula to calculate changes in parasite and host population in the following model:

$$\log N_{n+1} = \log N_n - \frac{aP}{2 \cdot 3} + \log F,$$

where the initial host density $(N_n) = 120$; the host rate of increase $(F) = 2$ and the initial parasite density $(P) = 50$.

Chapter 5

5.1   Table C shows the consequences of a population model to show how the presence of a density dependent mortality influences correlations. The data in the table below represent a hypothetical population over 15 generations. The population suffers two mortalities, (1) a density independent egg mortality (k_1) obtained from random tables, which represents the changing effects of weather, and (2) a strongly density dependent larval mortality (k_2).

Table C   Population model for exercise 5.1.

Generation	log Egg density $\log N_E$	egg mortality k_1	log larval density $\log N_L$	Larval mortality k_2	log adult density $\log N_A$
1	2·00	—	1·50	—	0·30
2	2·00	—	1·09	—	0·21
3	1·91	—	1·61	—	0·32
4	2·02	—	1·92	—	0·38
5	2·08	—	1·18	—	0·23
6	1·93	—	1·03	—	0·20
7	1·90	—	1·40	—	0·28
8	1·98	—	1·38	—	0·27
9	1·97	—	1·27	—	0·25
10	1·95	—	1·85	—	0·37
11	2·06	—	2·06	—	0·41
12	2·11	—	1·81	—	0·36
13	2·06	—	1·46	—	0·29
14	1·99	—	1·69	—	0·33
15	2·03	—	1·23	—	0·24

Plot the generation curves for larvae and for adults on a log scale and plot also graphs of the egg and larval mortalities (using k-values).

Questions   (a) What formula describes k_2 in the model?

(b) Why is the density independent k_1, which was introduced as a random factor, inversely correlated with both k_2 and log larval density?

Chapter 6

6.1 *Practical exercise.* Provide a sample of knapweed galls. The students can count surviving larvae and determine the causes of death of the others, identify the different parasite larvae, count and identify the egg shells of the various ectoparasites such as *Torymus*, *Habrocytus*, *Eurytoma robusta* etc. The results can be tabulated as in Table 6.1 and converted into the corresponding part of the life table for the host and for the parasites. Alternatively, any other suitable material can be used.

6.2 If no fresh material is available, take the bold figures from Table 6.2 or Table 6.3. Calculate for yourselves the remaining figures of these tables.

6.3 *The consequences of Nicholson's theory.* Given that the rate of increase of the knapweed gall-fly, *Urophora*, is 18, and the area of discovery of its parasite *Eurytoma* is $a = 0 \cdot 25$; calculate the steady densities of adult host and parasite where:

(i) *Eurytoma* is the only cause of *Urophora* mortality,

i.e.,
$$P = \frac{2 \cdot 3}{0 \cdot 25} \log \frac{N}{S} = ?$$

and
$$P = 17N \text{ therefore } N = ?;$$

(ii) 90 per cent of the *Urophora* larvae are killed before *Eurytoma* attacks;

(iii) 90 per cent of the *Urophora* larvae surviving the attack by *Eurytoma* are killed;

(iv) 90 per cent of the *Eurytoma* larvae are killed;

(v) 90 per cent of the *Urophora* surviving the attack by *Eurytoma* and 90 per cent of the *Eurytoma* larvae are killed.

6.4 Make a running calculation of eight successive generations for (v) [where there is 90 per cent mortality of both surviving *Urophora* and *Eurytoma* larvae] starting with initial densities of 6 adult *Urophora* and 2 adult *Eurytoma*. (a) In which generation does the population of *Urophora* reach its first minimum? (b) After how many generations is the parasite population less than a tenth of its initial value?

Chapter 7

Chapters 7 and 8 deal mainly with similar problems concerning insects with a single generation a year. The more elementary exercises for Chapter 7 use figures derived from models to show how well the methods work when we know the answer. The advanced exercises for Chapter 8 follow logically upon these and some of them use field data. Analysis is slow using a slide rule but quite quick if a small computer is available.

7.1 The derivation of a simple life table from census figures needs a moderate amount of arithmetic.

Population numbers are given in Table D from which life tables can be constructed for the host and for the parasite.

Table D The figures in each column represent population counts taken successively. Adult parasites are counted first. When they have attacked the host larvae these are counted and the number of them containing a single parasite larva is recorded. The number of adult hosts which emerge late in the year is the final figure for each year. Numbers are population density per m^2.

Year	0	1	2	3	4	5	6	7	8
Parasite adults P_A	—	1	1	3	6	15	47	41	4
Host larvae N_L	—	53	532	296	377	796	322	11	11
Parasite larvae P_L	—	3	24	45	121	491	306	10	3
Host adult N_A	15	12	56	34	32	29	2	0·4	—

Host figures Inspection shows that the rate of increase from adults to larvae is always well under 100, so 100 is a convenient notional figure to assume for the potential rate of increase. Calculate and tabulate the following:

1 host egg numbers N_E, assuming 100 eggs per adult, tabulate log N_E = log $N_A + 2$;

2 host larval numbers, converted to logarithms (log N_L);

3 mortality between egg and larval stages $k_1 = \log N_E - \log N_L$;

4 host larvae surviving parasitism $N_L - P_L$ and tabulate log $(N_L - P_L)$ = log S;

5 host mortality from parasitism, $k_2 = \log N_L - \log S$;

6 mortality of survivors after parasite attack up to adult stage $k_3 = \log S - \log N_{A2}$ where N_{A2} represents the adult population which produces the next generation;

7 find the generation mortality $K = k_1 + k_2 + k_3$. A check on your figures is prudent. Check that log $N_{A1} + 2 - K = \log N_{A2}$.

Key factor analysis: plot graphs of K, k_1, k_2 and k_3 against generation number.

Questions (a) Is the key factor k_1, k_2 or k_3?

(b) Do these graphs show any sign of density dependence—does one of the k-values change in the opposite direction to the key factor?

Testing for density dependence: for k_1, k_2 and k_3 plot the k-value for each mortality against the host density on which it acted.

Questions (c) Which of the k-values is density dependent?

(d) What formula best describes this k-value?

Note: the statistical method given in exercise 8.2.5 should be used as a final test of any apparently density dependent k-value found by the method given above.

Parasite figures

8 Calculate the value of the area of discovery

$$a = \frac{2 \cdot 3}{P} k_2.$$

Question (e) What form of model (see Chapter 4) describes parasite behaviour?

9 Estimate the mortality between the larval stage P_L and the resulting number of adults P_A in the next year the k value $k_P = \log P_L - \log P_A$.

Question (f) Is k_P more closely related to P_L or to N_L?

7.2 In this exercise the life table data for the host are already provided in Table E. The basic analysis of the figures can be done graphically in only a few minutes, because host populations are given as logarithms and the successive k-values for the different mortality factors are also printed.

Table E Host with synchronized parasite and three other mortality factors acting in succession.

Generation		1	2	3	4	5	6	7	8
Log egg population	log N_E	3·80	3·86	3·89	3·40	3·82	3·75	3·86	3·35
Juvenile mortality	k_1	0·5	0·3	1·2	0·3	0·8	0·4	1·4	0·2
Log small larvae	log N_{L1}	3·30	3·56	2·69	3·10	3·02	3·35	2·46	3·15
Adult parasites	P_A	250	292	473	115	173	168	279	70
Host larval mortality from parasitism	k_2	0·47	0·51	0·65	0·32	0·39	0·39	0·5	0·25
Log surviving host larvae	log N_{L2}	2·83	3·05	2·04	2·78	2·63	2·96	1·96	2·90
Larval mortality	k_3	0·73	0·82	0·42	0·71	0·65	0·78	0·38	0·76
Log host pupae	log N_p	2·10	2·23	1·62	2·07	1·98	2·18	1·58	2·14
Pupal mortality	k_4	0·04	0·14	0·02	0·05	0·03	0·12	0·03	0·08
Log host adults	log N_A	2·06	2·09	1·60	2·02	1·95	2·06	1·55	2·06
Generation mortality	K	1·74	1·77	2·29	1·38	1·87	1·69	2·31	1·29

1 Plot k-values on the ordinate against generation number on the abscissa. Put in the points for K, k_1, k_2, k_3 and k_4.

Question (a) Which is the key factor?

2 Tests for density dependency.

A. Draw scatter diagrams for k_1, k_2, k_3 and k_4 against the log of the population density on which they acted.

Question (b) Which k-values appear to be density dependent? Note that because k-values are measured by the difference between the logs of two successive population estimates, this test can give misleading results: k_1 in fact entered the model as a series of randomly chosen numbers, whose mean was $0 \cdot 5$. Yet a regression of k_1 on log N_E has a slope near $b = 1$, suggesting perfect compensation. Spurious results like this can arise from a number of causes; e.g., in a short series of random numbers successive values may be by chance negatively correlated. Other difficulties of testing for density dependence are discussed in Exercise 8.6 and the method given there should be used to test any plot of a k-value against log N which appears to be density dependent.

Question (c) What formula describes the relationship between k_3 and the log population density on which it acted?

Use the density dependent test given in Exercise 8.25.

Question (d) What formula describes the relationship between log S and log N between which k_3 acts?

Question (e) Do any other of these graphs suggest a simple relationship?

From Table E derive values for the area of discovery of the parasite. Use the formula derived from formula (4.4) that

$$a = \frac{2 \cdot 3 \, k_2}{P}.$$

Plot log a against log P.

Question (f) What formula best describes the relationship between log a and log P? Note that there are difficulties concerning accepting this relationship as statistically valid. Tests given in Exercise 8.2.6. should be used on any such apparent relationship.

Chapter 8

8.1 In Chapter 8 we tried to draw some conclusions from census figures which were in many ways incomplete. To demonstrate how our understanding of ecological processes is diminished if we do not have complete life table figures, reanalyse the model in 7.1 using only part of the data.

Question (a) Most field studies have no figures for P_A. In 7.1 what is lost if k_2 is measured but the measure of P_A is lacking?

(b) How might our interpretation be affected if we had no measure of parasitism? i.e., the successive measurements are of N_L and N_{A2} so that the mortality between these stages is $k_2 + k_3$.

8.2 An advanced exercise in the analysis of real life table figures.

Table F summarizes the life table data for winter moth from 1950 to 1968. The figures are rounded off to two (or occasionally three) decimal places. Use these data in the following exercises.

8.2.1 Calculate the densities of emerging adult winter moth (log adults = log larvae minus the sum of k_2 to k_6) and use these figures to calculate the egg densities. Assume (i) that the sex ratio is equality,

(ii) that 3/4 of the females bypass the traps and lay eggs, and

(iii) that each female lays 150 eggs.

A short cut is to assume that each **Adult** lays $3/8 \times 150$ eggs. The log of this added to the log adults = log eggs. As a check, log larvae $+ k_1 =$ log eggs. Plot the values of k_1 against the egg densities for the years 1956–1968.

Table F Summary of the life tables for winter moth

	1950	1951	1952	1953	1954	1955	1956	1957	1958	1959	1960	1961	1962	1963	1964	1965	1966	1967	1968
k_1	1·59	0·56	1·14	1·33	0·24	1·16	0·83	0·48	0·83	1·15	1·21	0·91	1·09	0·85	0·66	0·52	1·42	1·20	1·19
Log. larvae	2·05	2·07	1·74	1·26	2·20	1·89	1·98	2·44	2·28	1·76	1·33	0·88	1·13	1·61	2·12	2·43	1·71	0·99	1·00
k_2	0·11	0·01	0·003	0·01	0·004	0·003	0·03	0·02	0·07	0·10	0·02	—	0·02	—	0·005	0·03	0·04	0·01	—
k_3	0·04	0·02	0·06	0·02	0·03	0·03	0·01	0·01	0·02	0·04	0·06	0·06	0·06	0·04	0·01	0·01	0·005	0·03	0·01
k_4	0·05	0·03	0·05	0·04	0·06	0·04	0·02	0·05	0·04	0·02	0·03	0·04	0·03	0·01	0·01	0·01	0·02	—	0·03
k_5	0·69	0·77	0·75	0·34	0·73	0·65	0·47	0·79	0·82	0·66	0·72	0·22	0·24	0·40	0·74	0·87	0·86	0·50	0·48
k_6	0·29	0·10	0·03	0·16	0·07	0·09	0·28	0·20	0·16	0·15	0·45	0·08	0·06	0·12	0·14	0·13	0·33	—	—
Log. *Cyzenis* adults emerging	0·86	0·20	1̄·0	1̄·0	1̄·0	1̄·48	1̄·48	1̄·90	0·15	0·40	1̄·90	1̄·0	1̄·0	1̄·0	—	1̄·0	—	—	—
Log. *Cratichneumon* adults emerging	0·94	0·86	0·56	1̄·78	0·34	0·56	0·42	1̄·13	1̄·14	0·81	0·41	0·30	1̄·78	1̄·90	0·56	0·92	0·50	—	—

Question (a) Is the conclusion from the analysis of the years 1950–1962 that this mortality is density independent still valid?

8.2.2 For the years 1963–1968 plot the value of k_5 against the log densities of larvae falling into the traps, and thus continue the graph in Fig. 7.4.

Questions (a) Is the regression line still a good description of these data?

(b) Is the delayed component shown in Fig. 7.6 still evident?

8.2.3 The table gives the log densities of *Cyzenis* adults emerging to attack winter moth larvae, and the mortality (k_2) they cause. Similarly it gives the log densities of *Cratichneumon* adults and k_4 values. Use formula (4.4) that k-value for parasitism =

$$\frac{area\ of\ discovery \times parasite\ density}{2 \cdot 3}$$

to calculate the area of discovery for these parasites for each year.

Question (a) Do these agree better with the Nicholsonian or with the Quest theories of parasite action described in Chapter 4?

8.2.4 The number of larvae of winter moth per m² parasitized by *Cyzenis* can be found by subtracting the antilog of the log number surviving *Cyzenis* attack from the antilog of the log number of winter moth larvae before *Cyzenis* attacks. The table gives log numbers of *Cyzenis* adults emerging. Calculate the k-values for the mortality of *Cyzenis* between the larval and adult stages.

Questions (a) Is this mortality related to winter moth larval density?

(b) Is it related to *Cyzenis* larval density?

8.2.5 *Questions involving statistical tests*

Density dependence.

Explanation of the problems

To be valid regression analyses must use independent measurements. We must use log densities before (N) and after (S) the action of the mortality. However, normal regression methods assume that N is accurately known and that all the errors are in S; this is not true for most field measurements of density where N also has sampling errors. When some estimates or assumptions about the relative sizes of the errors in N and S can be made, there are statistical methods for tackling this sort of problem (see Snedecor 1967, and Bartlett 1949). But if these errors are unknown or cannot be estimated, the procedure shown below overcomes the difficulty using methods to be found in most standard textbooks of statistics. Use the densities of winter moth before and after the action of k_5 and make two tests.

(1) Make a correlation and regression analysis of log S on log N. Calculate the correlation coefficient r.

Question (a) Is the relationship significant? If so, calculate the regression coefficient b.

(2) Now, assuming all the errors to be in N and none in S, calculate the regression coefficient b for the regression of log N on log S. The correlation coefficient is the same.

Questions (b) Are the two regression lines on opposite sides of the line $b = 1$? If so a density dependent relationship has not been shown.

(c) Are the regression lines on the same side of the line $b = 1$?

(d) Are both lines significantly different from $b = 1$? If so, a formal proof of density dependency has been made.

Note that when an apparent density dependent relationship is seen from a plot of k-value on log N where these values are obtained from a model, as will be the case in exercises 7.2, the full test is not necessary. The question of errors in the measurement of log N does not arise. The only test necessary is to show that a regression and correlation analysis of log S on log N gives a significant slope which is also significantly different from a slope of $b = 1$.

8.2.6 *Testing the log a/log P relationship*

Explanation of the problem

From the formula (4.3) used to calculate the area of discovery

$$\left(a = \frac{1}{P} \log_e \frac{N}{S} \right)$$

it is clear that P is used in the calculation of a, and thus a regression analysis of a on P is not valid and independent measurements must be used to test the relationship. In this case the independent measurements are parasite density and parasitism represented by its k-value. Use the values of k_2 and the densities of *Cyzenis* adults and/or the values of k_6 and *Cratichneumon* densities and carry out the following analysis:

make a correlation and regression analysis of log k-value on log parasite density; calculate r.

Question (a) Is the correlation significant? If so, calculate b.

Because there are errors in the estimate of parasite density, calculate b for the regression of log P on the log k-values.

Question (b) Are both regression lines on the same side of (below) $b = 1$? If so, are both significantly different from the slope $b = 1$? If so, the log a/log P relationship has been proved. (Note, the regression line $b = 1$ is the one on which points will fall if the area of discovery is a constant.)

Note that, as in the previous exercise, when a model is being analysed the question of the effect of errors in the measurement of P does not arise. The only test necessary is to show that a correlation and regression analysis of the log k-value for parasitism on log parasite density is both significant and significantly different from a slope of $b = 1$.

8.3 Analyse the data for the black headed budworm given in Table 8.1.

(1) Convert the numbers in line 2 to logarithms and tabulate log N.

(2) Convert the percentages of parasitism to the corresponding k-values, k_p.

(3) Assume a 10-fold potential increase (add 1 to log N). The generation mortality $K = 1 + \log N_n - \log N_{n+1}$. Tabulate the k-values for the residual mortality $k_r = K - k_p$.

(4) For a key factor analysis plot K, k_p and k_r against generation number ... your graph should agree with Fig. 8.6.

(5) Plot k_P against log N.

Question (a) Is this a delayed density dependent relationship?

(6) Plot the values of k_r for each generation against the log survivors from parasitism ($\log S = \log N - k_P$).

Question (b) Is this a delayed density dependent mortality?

Chapter 9

The changes from generation to generation in a pest population like that of the winter moth can be represented by the following model:

$$\log N_{Ln+1} = \log N_{Ln} - k_2 - k_3 + \log F - k_1,$$

where N_{Ln} and N_{Ln+1} are the pest populations in successive generations. In this particular model, the potential for increase, $F = 75$; k_1 is a density independent mortality which can be taken as a constant $= 0 \cdot 7$;

$$k_2 = \frac{0 \cdot 056\, P_n^{1-0 \cdot 52}}{2 \cdot 3}$$

is the mortality caused by P_n searching parasites from the formula (4.12) where $m = 0 \cdot 52$ and $Q = 0 \cdot 056$. The number of pests parasitized minus a mortality due to k_3 gives the next generation of searching parasites P_{n+1}.

$k_3 = 0 \cdot 35 \log N_n$ and is a density dependent mortality caused by predators which kill pests and parasite larvae indiscriminately. Use the above data to calculate successive generations until the answers to the questions below become evident; commence with a pest density of 1000 and a parasite population of 30.

Question (a) What is the steady density of the pest N_L? Starting from the steady density of host and parasite how would the following insecticide treatments affect the pest N_L population?

(b) A 70 per cent kill of the pest acting just prior to parasitism in each generation.

(c) A 70 per cent kill as in (b) which also kills 70 per cent of the searching parasites in each generation, and

(d) A 70 per cent kill as in (b) which also completely eliminates the predators.

(e) What part of the natural enemy complex should be preserved in any control programme?

ANSWERS TO QUESTIONS IN THE EXERCISES

1.2   (a) Egg and first instar, (b) 1st–2nd instar, (c) pupa–adult, (d) $0 \cdot 37$.

2.2   (a) The consequences cannot be determined until the rate of increase is known.

2.3   Yes.

2.4   b is correct.

2.5   d is correct.

2.6   (a) Between 2 and 3, (b) extinction follows if F is more than 4 or below 1.

2.8   (a) The result is in Fig. 3.10.

2.9   (a) At higher densities the effect is more severe than 'contest', (b) $0 \cdot 05$, (c) $1 \cdot 2$, (d) $25/477 = 0 \cdot 052$.

3.2   (a) Yes, (b) yes.

4.1   (a) No, (b) yes, (c) yes.

4.2   The value of area of discovery falls with N because of the large handling time.

4.3   (c) $\log a = -0 \cdot 76 - 0 \cdot 68 \log P$.

5.1   (a) $k_2 = 0 \cdot 8 \log N_L$, (b) the egg density varies very little because k_2 has almost stabilized the adult population; the variable k_1 therefore largely determines larval density which in turn determines k_2.

6.3

	(i)	(ii)	(iii)	(iv)	(v)
P	$11 \cdot 6$	$2 \cdot 36$	$2 \cdot 36$	$11 \cdot 6$	$2 \cdot 36$
N	$0 \cdot 68$	$0 \cdot 295$	$0 \cdot 295$	$6 \cdot 8$	$2 \cdot 95$

6.4   (a) 4, (b) 6.

7.1   (a) k_1 is clearly the key factor at first, but by year 6 parasitism (k_2) is even more important.
(b) k_3 always changes in the opposite sense to the change in k_1.
(c) k_3 increases with $\log N_L$.
(d) In fact a better description is $k_3 = 0 \cdot 4 \log S$. This is because k_3 acts on the total larval population, whether parasitized or not.
(e) Using the figures given in the Table the area of discovery varies between $0 \cdot 5$ and $0 \cdot 7$. It was put into the model as a constant—the variation was introduced when the parasite numbers were rounded off to whole numbers.
(f) The mortality of the parasite k_P is related to host population $\log N_L$ ($k_P = k_3$).

7.2   (a) The key factor is k_1, (b) k_3, (c) $k_3 = 0 \cdot 4 [(\log N_{L2}) - 1]$, (d) the slope $b = 1$ minus the slope found in (c), (e) no—but k_1 might have a density dependent component. It needs testing by the methods in 8.6. (f) $\log a = -1.16 - 0.5 \log P$

8.1   (a) The delayed density dependent effect still shows when k_2 is plotted against log N_L, but without values for P_A we cannot model k_2.
       (b) k_3 is density dependent, but there is no significant slope in the regression slope for $(k_2 + k_3)$ against N.

8.2.1 (a) It shows a delayed density dependent relationship.

8.2.2 (a) Yes, (b) yes.

8.2.3 (a) Quest.

8.2.4 (a) No, (b) yes.

8.2.5 (a) Yes, (b) no, (c) yes, (d) yes.

8.2.6 (a) Significant in both cases, (b) no.

8.3   (a) Yes—the points when joined consecutively form a spiral, (b) yes.

9.1   (a) Log $N_L = 3 \cdot 13$, (b) log $N_L = 2 \cdot 35$, (c) log $N_L = 2 \cdot 37$, (d) the population rises at first and oscillates with decreasing amplitude, (e) predators if they act as density dependent factors.

a   Area of discovery. The total area effectively searched for hosts by a parasite throughout its lifetime (p. 59).

a'   Coefficient of attack. An instantaneous rate of encountering hosts. It becomes the area of discovery when multiplied by searching time (p. 68).

α; β Interspecific competition coefficients (p. 39).

Alternations   The population density alternates between high and low values in successive generations if the key factor is density dependent and strongly over-compensates for change in population density (Fig. 2.10, C, D, E).

'Competition curve'   The relationship between percent parasitism and the area traversed by a parasite population Pa (after Nicholson 1933). The curve rises asymptotically towards 100% as parasite density increases (Fig. 4.2 D).

C   Average egg complement of a female parasite (p. 57).

Competitive exclusion   The elimination of all but one of competing species (p. 37).

Congeneric species   Species belonging to the same genus.

'Contest' competition   Where each successful competitor gets all it requires for survival or reproduction—the remainder get none or insufficient. (p. 25).

Density dependent   A proportionate increase in mortality (or decrease in fecundity) as population density increases (p. 18, Fig. 2.6 A).

Density independent   The percentage mortality or the survival varies independently of population density (p. 18, Fig. 2.6 C, D).

Delayed density dependent   A parasite will act as a delayed density dependent mortality factor on the host if its rate of increase is strongly correlated with host density in successive generations, as is assumed in Nicholson's theory (p. 65).

Developmental zero   The temperature T_0 at which a given developmental process would cease if the rate of the process was proportional to $(T-T_0)$. (Fig 5.2).

Diapause   A physiological condition in which development is arrested, often at a particular stage, which tides the insect over an unfavourable season.

F   Reproductive rate.

Generation curve   The population density of a given developmental stage plotted against generation number for a sequence of generations (Fig. 1.1 c).

Inverse density dependence   A proportionate decrease in mortality (or increase in fecundity) as population density increases (p. 18, Fig. 2.6 B).

k   *k*-values ($k = \log N - \log S$). Subscripts (e.g. k_0, k_1, $k_2 \ldots k_n$) refer to successive stages at which the mortalities act.

K   Generation mortality expressed as a *k*-value.

κ Carrying capacity. The equilibrium population density from the logistic equation (Fig. 2.2).

Life table A description of the age-specific survival of cohorts of individuals in relation to their age or stage of development (Chap. 6).

Logarithmic reproduction curve A reproduction curve with logarithmic axes (p. 22, Fig. 2.9).

Logistic equation (= Verhulst–Pearl equation) (p. 13).

$$\frac{dN}{dt} = r_m N \left(\frac{\kappa - N}{\kappa} \right)$$

m Mutual interference constant. A measure of the rate of decrease in searching efficiency as parasite density increases (p. 69).

N Number of hosts.

N_1; N_2 Population densities of species 1 and 2.

N_a Number of hosts encountered by P searching parasites.

N_A Number of adults.

N_E Number of eggs.

N_{ha} Number of hosts attacked and successfully parasitized.

N_L Number of larvae.

N_n; N_{n+1} Population density in generations n and $n+1$.

N_P Number of pupae.

N_s Equilibrium density (= 'steady density') of hosts.

N_0; N_t Population density at time 0 and time t.

Oscillations Regular cyclic population changes which are the consequence of delayed density dependent factors, and have peaks in the generation curve which are 5 or more generations apart (Figs 4.3, 4.9).

P Number of searching adult parasites.

P_E Number of eggs laid by P parasites.

P_s Equilibrium density (= 'steady density') of parasites.

Partial population curve The population density of a given developmental stage plotted against time (Fig. 1.1).

Q Quest constant. The area of discovery when parasite density is 1 (p. 69).

r_m The intrinsic rate of natural increase (p. 12).

Reproduction curve The relationship between the numbers of a given stage in generation $(n+1)$ plotted against the numbers of that stage in generation (n) (p. 20, Fig. 2.7 c).

S Number of hosts surviving parasitism.

'Scramble' competition Where the resource is shared equally amongst the competitors (p. 25).

Successive percentage mortality (= 'apparent mortality') The mortality in each developmental stage expressed as a percentage of the number alive at the beginning of the stage.

T_h Handling time (p. 68).

T Total time initially available for searching (p. 68).

Total population curve The total population density of individuals of all stages plotted against time (Fig. 1.1).

INDEX

209